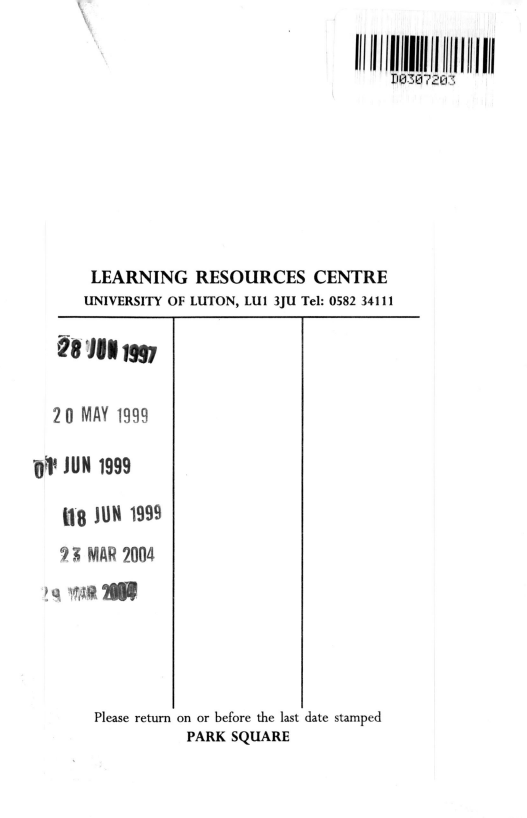

IEE TELECOMMUNICATIONS SERIES 33

Series Editors: Professor J. E. Flood
Professor C. J. Hughes
Professor J. D. Parsons

MODERN PERSONAL RADIO SYSTEMS

Other volumes in this series:

MODERN
PERSONAL
RADIO
SYSTEMS

Edited by

R. C. V. MACARIO

THE INSTITUTION OF ELECTRICAL ENGINEERS

Published by: The Institution of Electrical Engineers, London,
United Kingdom

The Institution of Electrical Engineers,
Michael Faraday House,
Six Hills Way, Stevenage,
Herts. SG1 2AY, United Kingdom

British Library Cataloguing in Publication Data

A CIP catalogue record for this book
is available from the British Library

ISBN 0 85296 861 2

Printed in England by Short Run Press Ltd., Exeter

Contents

Preface

The earlier textbook in the IEE Telecommunications Series, *Personal & Mobile Radio Systems*, published in 1991, has proved very popular and has had to be reprinted several times. The contents of this earlier book have not dated, and still provide an introduction to the fundamental aspects of personal radio-telephone technology.

However, it became clear to the Series Editors and to those of us involved in teaching, training, development, management and operation of personal communication systems that a new text was needed. We decided again to bring together expert contributions and combine them into a uniform and easy to read text.

The result is a book that takes over where the previous book had to leave off, partly because the knowledge and standards were not yet developed and available when it was written.

I have again been most fortunate in that I have been able to call on many of the key people associated with the technology and strategies of modern personal radio systems. I am very appreciative of the extremely positive text received for each chapter, and of the encouragement I have received in fashioning the contributions into a uniform text.

As with the earlier book, I feel that the list of contributors shows that we have tried to prepare a text evenly distributed across many centres of research, development and operation involved in the personal communication business. Many other companies and colleagues have also contributed their advice on one topic or another, and I am very grateful to many for their interest and advice.

Much of the text preparation was undertaken by Ruth Baker: a far from simple task, because most of the contributions arrived in different styles and written on incompatible word processors. I would like to record my appreciation of her devotion to the project. The IEE publishing department at Michael Faraday House, particularly Fiona MacDonald, also provided very positive support and feedback for the book, which sees the introduction of coloured diagrams in places.

R. C. V. MACARIO
University of Wales, Swansea
December 1995

Contributors

David E.A. Britland
APD Communications Ltd.
Ardenham Court
Oxford Road, AYLESBURY
Bucks. HP19 3EQ
01296 435831

Dr. Alastair Brydon
Cellnet
260 Bath Road
SLOUGH SL1 4DX
01753 565697

Dr. Alister Burr
Dept. of Electronics
University of York
Heslington
YORK YO1 5DD
01904 432335

Simon Cassia
Motorola Ltd.
Land Mobile Product Sector
Jays Close
Viables Industrial Estate
BASINGSTOKE
Hants. RG22 4PD
01256 484363

Dr. Stanley Chia
Cellular Systems
MLB G 53C
BT Laboratories
Martlesham Heath
IPSWICH IP5 7RE
01473 643416

Bruce R. Elbert
Hughes Communications, Inc.
P O Box 92424
LOS ANGELES CA90275
U.S.A.
001 310 607 4132

Vernon Fernandes
Motorola Ltd.
European Cellular
          Infrastructure Div.
16 Euroway
Blagrove
SWINDON SN5 8YQ
01793 541541

Ian Harris
Vodata Ltd.
Elizabeth House
13/19 London Road
NEWBURY RG13 1JL
01635 503270

Dr. Ray Macario
Dept. of Electrical Engineering
University of Wales, Swansea
Singleton Park
SWANSEA SA2 8PP
01792 295416

John W. Mahoney
Cellnet
260 Bath Road
SLOUGH SL1 4DX
01753 565573

Dr. Amelia Platt
Science and Engineering Research
    Centre
The Gateway
De Montfort University
LEICESTER LE1 9BH
01162 577586

Dr. Peter Ramsdale
One-2-One
Imperial Place
Maxwell Road
BOREHAMWOOD
Herts. WD6 1EA
0181 2142070

Professor Raymond Steele
Multiple Access Communications Ltd.
Epsilon House
Chilworth Research Centre
SOUTHAMPTON SO16 7NS
01703 767808 / 592881

Dr. Walter Tuttlebee
Roke Manor Research Ltd.
Roke Manor
ROMSEY
Hants. SO51 0ZN
01794 8335000

Chapter 1

Modern personal radio systems requirements and services

Peter A Ramsdale

Introduction

What is personal communications? Is it a cellular radio network, some combination of cordless telephones, paging, or a fixed network incorporating Intelligent Network (IN) technology? Rather than defining a technology, personal communications is best described by the features individual users would expect from their 'ideal' personal telecommunications service. Because of this individual nature a single definition is not possible, but the attributes sought by most people can in general be described under a common vision.

For example in January 1989, the UK Department of Trade & Industry published a consultative document entitled 'Phones on the move'. This embodied a vision of a new kind of telephone service available through the use of a single personal handset connected to a network by radio. It would offer a two-way, fully mobile service competing both with existing cellular and with public switched fixed network services. In December 1989, licences for Personal Communication Networks (PCNs) were awarded.

Since then, almost all participants in the mobile communications arena have tried to link their services to the personal concept. What was once known as 'mobile communications' has become 'personal communications'. Several mobile telecommunications services may be described as the precursors, rather than the providers, of personal communications, ranging from simple paging and early cordless telephony to analogue cellular and the first digital public cordless service, known as telepoint.

'Phones for people, not places' and 'the mobile phone for everyday, for everyone' are more than useful slogans. They are the keystone of the PCN philosophy, to treat people as individuals and align telecommunications to suit specific lifestyles. Providing greater choice for the customer was one of the driving forces behind the creation of the licences. PCN will fulfil this brief not just by providing more sophisticated telephone services for the person on the move, but by tailoring service packages to meet the demands of a wide variety of users, ranging from the business executive, the doctor or the builder to the teenager, the working parent or the house bound senior citizen. This is the real implication of the personal communications vision.

The vision of personal communications focuses on the provision of high quality two-way telecommunication services - speech and data - to both business users and consumers on the move, outdoors and indoors. The requirements of personal communications are essentially a service concept rather than a radical change in network technology. However, advanced technical solutions are needed to deliver a high-capacity service of speech, data and supplementary services at a high quality level from a competitive cost base. Such solutions enable personal communication systems to replace services in both fixed and mobile markets. PCN is the first system attempting to bring both fixed and mobile applications together; a personal service from one handset, not a separate mobile and fixed phone.

1.1   Features of personal communications

Personal communication systems aim to provide their users with telecommunications services throughout specified coverage areas. For some users, mobility is the key feature provided by the system, but for mass market acceptability any shortfall in service compared with the fixed PSTN, becomes increasingly important.

A fixed PSTN can provide an element of personal mobility in that a user can be reached at any termination point of the network. A radio terminal introduces additional mobility either by providing cordless coverage regions around network termination points or total movement across the cells of a cellular radio network.

In radio based systems it is important to use the limited radio spectrum as fully as possible and figures of merit are defined to measure the frequency reuse. A typical measure of efficiency is traffic (in Erlangs) /MHz/ square km. High spectral efficiency relies on utilising small cells, reusing frequencies in nearby cells, etc.

Personal radio terminals should have strong user appeal, as this is the customer's tangible means of contact to the service. Small size (weight and volume), long battery life and low cost are clear requirements. Other areas of customer concern include consistency of radio coverage, speech quality, cost of usage, call security and access to advanced telephony features.

1.2   Radio communication systems

The way in which radio communication systems develop depends on many factors such as:

Regulation
Radio spectrum allocation
Standards
Technology
User requirements
Service positioning
Investment

At various times, different combinations of these factors will drive change.
Sometimes a key technical development, such as changing from analogue to digital
technology, is most significant, while at other moments a market may take off
through the timed and heavily promoted introduction of a new tariff package.
Some of these drivers of change will be considered further.

1.2.1 Regulation

There is an increasing liberalisation of the telecommunications market throughout
the world with the aim of eliminating monopolies and introducing more
competition, but this extra competition has to be carefully introduced if it is to be
effective. For example, to aid new operators, favourable start-up conditions may
be required in order to compete with well established operators.

A further regulatory requirement, in a competitive environment, is to prevent
cross-subsidies between different functions within one company. For example:

- independent service providers should be given the same terms as a service
  provider group within the operator's company,

- at defined open-interfaces, the same rates should be charged, whether they
  be related or unrelated operators.

An environment, which is becoming increasingly common, is that of multiple
operators who compete, but must also co-operate in delivering service to
customers. Often a telephone call from a cellular radio based network will be
carried over a leased line of another operator and finally delivered to a telephone
on the network of a fixed PSTN operator.

Currently, the main regulatory developments are related to operator licences
and inter-network access. For example, a Commission of the European
Community green paper on 'Mobile Telecommunications' has recognised that
operator licences, tied to a particular technology, are a 'barrier to overcome'. This
approach gives the maximum flexibility for building a service from the underlying
technologies. Open Network Provision (ONP) directives are likely to be
implemented in 1998. This will help European countries to establish effective
competition between operators and public telecommunication networks.

1.2.2   Radio spectrum allocation

There is worldwide agreement on the general class of use of radio spectrum in all frequency bands, i.e. fixed, mobile, space, etc. National administrations licence operators to use blocks within this framework for specific applications. The amount of spectrum allocated clearly critically affects the service which can be supported. In cellular radio, spectrum is clearly regularly reused, but there are limits to the number of cells which can realistically be deployed, and hence a limit on the total capacity of the network.

A further cost driver is the method used for spectrum pricing. This may be a simple licence cost, based on the administration costs of the radio regulator, or a greater cost aligned to some view of spectrum worth, or a cost realised at an auction of spectrum to potential operators.

High spectrum costs can affect the solution chosen for a particular application. An anomaly could be created, for example, by an operator paying a large fee for spectrum for cellular radio, while a cordless telephone occupies 'free' spectrum in an *uncontrolled* band. Hence a dual mode (cordless/cellular) handset would allow a mobile user to access fixed network tariffs when near enough to his cordless base station, but cellular rates elsewhere.

In the USA, radio spectrum has been auctioned for personal communication systems (PCSs) at 1900 MHz. The result has been that the operators are paying much more for this spectrum than their 800 MHz cellular allocations, or the radio licence fees paid by European operators. The need to amortise this cost across many customers encourages the spectrum to be used for a mass market proposition, but the approximate doubling of costs/customer, i.e. infrastructure plus licence fees, makes it harder to set tariffs which can compete with the local loop, an example of this being PCN in the UK.

1.2.3   Standards

Pan-European standards have been developed in order to benefit from the economies in scale of serving a larger market. These standards are produced by ETSI (European Telecommunications Standards Institute) whose membership includes administrations, manufacturers and operators from the whole of Europe. A particular success has been GSM, the pan-European digital cellular radio standard. This enables customers to roam across Europe and beyond without changing their handsets.

Although all parts of a telecommunications network can be standardised, it is the external interfaces and radio terminals which are most important. In Europe, the mutual recognition of radio terminals/handsets means that a handset, which is type approved in one country, satisfies the type approval requirements in other countries.

1.2.4 Technological developments

Standardisation is taking place against a background of rapid technological development. Today's fixed operators are adding mobility features, based on intelligent network technology, in addition to deploying wireless local loop extensions to their fixed infrastructure. Handset technology is rapidly evolving, with size, weight and ex-factory costs rapidly decreasing. In the next few years, multi-mode handsets will appear based on a combination of complementary but incompatible standards for cordless, cellular and satellite systems. Combined with intelligent network concepts an integrated telephone service will be possible.

New technology should continue to offer new capabilities and drive costs down. Both device and systems developments are important. It is expected that flexible bit rate air interfaces, high spectral efficiency/reuse, and data compression techniques, will be of particular relevance due to the demands for high capacity within a finite radio spectrum.

1.2.5 New services

The commercial viability of new personal communications services relies on the takeup of these services. To date, market research has confirmed that the majority of mass market mobile users simply want high quality, low cost voice telephony. At present, the data market for mobile and personal communications represents a niche market sector. By the end of the century, the size of the data market will increase considerably, but it is difficult to see the vast majority of users wanting to use the more exotic high bit rate services. Hence these services must be provided in such a way that the cost of basic telephony is not increased.

The ultimate goal could be to provide everything found on the fixed telephone, at no greater price, and with unrestricted mobility. However, this may well be limited only to services where a pocket terminal is a suitable man-machine interface; for example, A4 size fax may not be appropriate.

1.2.6 Infrastructure investments

Operators (both fixed and mobile) make huge investments in network infrastructure. Manufacturers have long pay back periods during which they must recoup the development costs of new switching, network and radio products. Similarly, owners of handsets have expectations that these will not become obsolete prematurely. These investments can lead to a period of at least ten years between revolutionary changes being made to networks. It also encourages operators to use an evolutionary approach so as to maximise the profitable lifetime of their network investment. A revolutionary change to infrastructure must deliver some significant advantage in order to be justifiable. Going forward, consideration is being given to migratory routes such that revolutionary advances are made in specific areas only and unnecessary charges are minimised.

1.3 Cellular radio concept

The key to mass market personal radio systems is frequency reuse. The cellular radio system concept is to use frequencies regularly in a grid of cells across the required coverage area. This is the concept of *frequency reuse* which refers to the use of radio channels, on the same carrier frequency, to cover different areas which are separated from one another by sufficient distances so that co-channel interference (C/I) is manageable.

Cell splitting refers to growth within a cell requiring the boundaries of an original cell to be revised, so that the area of the original cell now contains several cells.

Figure 1.1 shows an idealised regular cell plan with an N cell frequency reuse pattern, in this case N=3. A mobile at the edge of a cell is receiving a wanted signal at maximum range and interference from a 'ring' of six co-channel interferers. The C/I is equal to the wanted signal divided by the sum of these six interferers, the next ring being negligible.

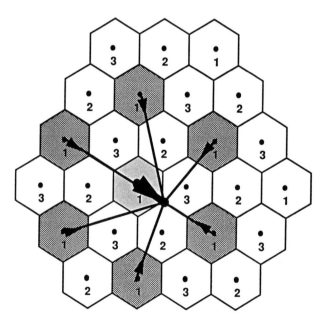

Figure 1.1 Co-channel interference

It is clear that C/I improves as N is increased, but because the number of channels/cell is equal to the total number of channels divided by N, the capacity of the network falls.

New digital standards for cellular radio use modulation and coding techniques which possibly tolerate lower values of C/I, so that frequency reuse can be improved. The performance of digital speech encoders under marginal C/I conditions is particularly important and defines the point at which speech quality declines significantly.

By looking at Figure 1.1, it can be seen how sectorisation improves C/I. In a tri-sectored cell, separate directional antennas are used for each 120° sector. Hence a mobile will only be exposed to three interfering signals thus allowing these to be brought closer and hence enabling a lower value to be used for N.

In a practical cellular radio network, it is not possible to acquire cell sites which fit exactly on a regular grid, and variable heights, and terrain features, complicate the situation. Hence, propagation modelling tools are used extensively to predict the interference levels between cells. However, the idealised grid gives guidance to the potential capacity and coverage performance levels across the network.

In a nation-wide cellular radio network there are two distinct regions:

- co-channel interference limited
- coverage area limited

Urban areas are generally C/I limited as due to the high user density, small cells are used.

Conversely, rural area cells are made as large as the propagation loss will allow, because the required channels/sq km is much lower.

Comparing 900 and 1800 MHz networks it is found that cell sizes are similar in urban areas, but in rural areas an extra 10 dB propagation loss for the latter halves the diameter of radio cells.

1.4 GSM cellular system

Cellular radio, originally developed in the USA during the 1970s, is known as the Advanced Mobile Phone System (AMPS) at 800 MHz. In Europe, several diverse standards were deployed including Nordic Mobile Telephone (NMT 450 and NMT 900) systems at 450 and 900 MHz respectively and in the UK, the Total Access Communication Systems (TACS), which is closely related to the AMPS standard.

In 1982 work commenced on a pan-European cellular radio system standard for Europe known as GSM. The key ideas behind GSM were that it would provide a compatible standard across Europe and permit roaming throughout the service area. Digital technology would be used so as to increase network capacity, provide higher quality and improve security. Costs would be reduced, due to advances in digital technology, and the high equipment volume and competitive sourcing opportunities afforded by a large market size.

GSM set out to be at least as good as existing analogue solutions in all parameters and better in some. The requirements of GSM include the following:

Services:

- Users should be able to call or be called across Europe.
- Advanced PSTN services should be provided consistent with ISDN services albeit at limited bit rates only.
- Handheld and vehicle mobiles should be supported.

Quality of service:

- The quality of service should be at least as good as 900 MHz analogue systems.
- Encryption should be used to improve security for both the operators and the customers, with little added cost.

Radio frequency utilisation:

- High spectrum efficiency should be achieved at reasonable cost (i.e. the standard should allow high frequency reuse).
- The system should operate in the paired frequency bands of 890-915 and 935-960 MHz and must be able to co-exist with earlier 900 MHz systems. This is essential for smooth migration of spectrum from analogue systems to the new standard.

Network aspects:

- CCITT identification and numbering plans should be used.
- An international signalling system should be utilised.
- There should be a choice of charging structure and rates.
- No modifications should be required to the PSTN due to its interconnection to GSM.
- Signalling and control information should be protected.

Cost:

- The system parameters should be chosen to limit costs, particularly mobiles and handsets.

The GSM standard meets these objectives and is being successfully deployed throughout Europe and beyond.

1.5   Personal communication networks

Cellular radio networks were initially planned to provide service for mobile car telephones. As portable handsets became smaller and their usefulness was appreciated, the networks improved coverage to provide indoor as well as outdoor capability. The networks were connected to the PSTN as a trunk interconnection giving a single tariff, which is quite appropriate for customers frequently moving across a wide area.

PCN, as the new personal system is known, first started in the UK through the government wanting new services to compete with the mobile cellular operators in the short term and the local loop, in the longer term. Subsequently other countries have awarded licences to create more competition.

The architecture of a PCN network is illustrated in Figure 1.2. This shows a set of subsystems and interfaces which are defined in detail in a series of ETSI recommendations [1]. The key features of these recommendations are the description of open interfaces to the OSI framework and the use of ISDN standards for signalling and network functions. The interface approach gives operators flexibility in equipment procurement. A brief description of the PCN now follows.

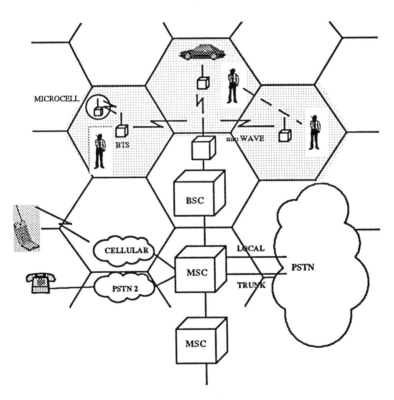

Figure 1.2 Personal communications network

The Base Station Controller (BSC) controls and manages a number of Base Transceiver Stations (BTSs) by providing the lower-level control of cellular functionality. ISDN signalling protocols are used within this Base Station Subsystem (BSS) and over the air interface.

The Mobile Switching Centre (MSC) is primarily concerned with routing calls to and from mobile stations. The Home Location Register (HLR) contains the customer information required for call routing and administration, including class of service data identifying to which services a particular customer is allowed access. Associated with each MSC is a Visitor Location Register (VLR) which stores information, including detailed location data, on all mobiles currently active within that MSC's area of control.

Network access is controlled by algorithms which carry out rigorous authentication of the mobile stations. The Equipment Identity Register (EIR) provides an up-to-date check on the validity of mobile equipment, while the Authentication Centre (AUC) checks the validity of the Subscriber Identity Module (SIM). A feature of GSM and PCN is that access to the network is granted through an SIM which is generally mounted on a smart card plugged into the mobile station to personalise the equipment for a particular customer. Security is provided by the use of challenge/response pairs of signals for authenticating access to the network and encryption keys for decoding enciphered speech, data, and other signalling passing over the air interface. All speech is digitally encoded at 13 kbps (full rate codec).

PCN has concentrated on handportable phones from the outset. The networks aim to provide high quality indoor and outdoor coverage requiring a large number of cells. These can only be justified by attracting large numbers of users such that mass market economies of scale are achieved. This helps to drive down the cost of both the handsets and the network. By connecting the network to the PSTN for every charge group and negotiating interconnect rates similar to those between fixed PSTN operators, local as well as trunk tariffs can be set.

To help such a service to emerge, 150 MHz of spectrum around 1800 MHz was made available for three operators in the UK (although subsequently only two reached service). This gave the new operators a greater potential network capacity than the two existing cellular companies, particularly in cities. For rural areas with a much lower customer density, cells tend to be as large as radio propagation losses will permit and the greater propagation loss at 1800 MHz means that the cost of providing rural coverage is greater.

Further changes to help PCN operators were to allocate part of the 38 GHz spectrum for cost effective connection between cell sites and switching centres, to allow PCN operators to sell directly to customers, rather than through service providers, and to allow inter-network roaming in rural areas as part of the service coverage area obligation.

The operators were licensed in December 1989, and Mercury One-2-One launched in September 1993, followed by Hutchison Orange in April 1994. Customer growth has been rapid with Mercury One-2-One gaining 260,000 customers by the end of March 1995, for example.

1.6 PCN standard

To meet the requirements of the personal communications vision, the only suitable technology basis is that of cellular radio. Fixed networks can provide personal mobility, such that users can use any access point, but some form of radio access is needed for terminals to be portable. Cordless technologies can be used to provide radio links to access points on a fixed network. Examples of cordless telephones with defined air-interface access are CT2 and DECT, but these standards do not define routing, switching and service functions which must be provided by the service network. However, cellular networks define complete telecommunications networks, including full terminal mobility, such that handsets are continuously affiliated to the network and handover takes place from cell to cell with handsets maintaining calls even when highly mobile.

A further advantage of cellular rather than cordless technology, for PCN, is in the range of cell sizes. Cordless access technologies have only been designed for quite small cells (up to 200 m) and 'stretching' the cells is difficult. In the case of CT2, range is limited by fast fading as there is no error protection on the speech channel. For DECT, the data rate (bit period = 868 ns) was designed primarily for indoor operation where delay spreads are short (typically 50 ns) and DECT does not have channel equaliser for greater delays.

Cellular air-interfaces tend to be more complex and require more elaborate transceivers but these are suitable for a wide range of cell sizes, both indoors and outdoors, such that they can be used in small (high capacity) cells and large (wide coverage area) cells.

The DTI required UK PCN to be based on existing standards, produced by ETSI, as this would enable early introduction of the service and encourage harmonisation across Europe. In their PCN licence applications to the DTI, the successful applicants proposed adapting the 900 MHz GSM standard for operation in the 1.8 GHz band, subsequently known as DCS 1800 [2].

GSM was chosen because it provides a good match to PCN requirements, but at the same time, it was recognised that enhancements to the GSM standard were needed not only to translate its operation to the 1.8 GHz band, but to refine the standard in order to:
- support small, low power handsets (1 watt)
- allow close proximity working
- access the wider available frequency band, 150 MHz compared with 50 MHz for GSM.

Another enhancement has been to support a multi-operator environment by incorporating 'national roaming'. 'International roaming' as defined by GSM caters for mobiles belonging to a network visiting another country and receiving service from a network of that country. National roaming allows operators of an individual country to restrict roaming to selected areas of the visited network. The key feature of the standard is to prevent a mobile remaining on the visited network when the home network is available.

A further development enhances the addition of microcells to a network [3]. The problem is the risk of dropping calls from mobiles when they leave the microcell coverage area, for example at street corners. The enhancement to the standard permits stationary and slow-moving handsets to access the microcell but ensures that fast-moving mobiles remain served by conventional macrocells.

DCS 1800 has provided telecommunication operators with an opportunity for the initial step into the PCN marketplace. The standard represents the technical foundation for the implementation of the network and as such it provides for a wide range of options and design variants suitable for PCN. However, the specific network design and implementation are equally important in determining the quality and type of service actually offered to customers.

1.7 Implementation of PCN

Although the elements of a mobile cellular network and PCN are similar, the ways in which the networks are implemented can vary significantly in order to match their different market objectives. Traditionally cellular networks were initially deployed with as few cells as possible to provide coverage just over areas frequently visited by mobiles, e.g. city centres, motorway corridors, etc. The tariffing has little call distance structure and comparatively high charges and is positioned as a premium service ideal for those requiring wide-area mobility. As the business develops, enhancement to the coverage in terms of coverage quality and capacity can be financed by the revenue from the existing customers. The better coverage increases suitability for hand portable users. Eventually the network can be used for other types of customer, for example, someone desiring an occasional emergency service can be offered a package with lower ownership charges and very high usage charges, because the network build costs have already been paid for by the principal users.

PCN operators seek to offer high quality communications with a customer base requiring contiguous coverage over the 'community of interest' where the service is being sold. In general this community is at least a city plus commuter environs covering hundreds (possibly thousands) of square km. PCN requires a very large initial investment to ensure that the network quality meets the marketing requirement from launch. By building the 'final' network at the outset, the total costs are less due to economies of scale in manufacturing, deployment etc. together with savings from not requiring expensive upgrading, cell splitting, cell replacement, re-planning etc. commonly experienced in evolving cellular networks. However as none of the initial network can be financed by revenue from users, PCN requires long term investment commitment and recognition of the considerable period before profitability.

1.7.1   Short message service

The GSM/DCS 1800 standard provides a structure for delivering a full range of telecommunications voice and data services using modern signalling and control structures. In addition to this, the Short Message Service (SMS) is a feature which provides for delivery of messages of up to 160 characters both to and from the mobile in a connectionless manner, i.e. no speech path set up required. SMS may be delivered both to 'addressed' mobiles (point-to-point service), or on a general broadcast basis from individual, or groups of base stations (cell broadcast mode). This latter mode is particularly useful for general or localised information services.

In the PCN environment, messaging (both voice and data) can provide a very powerful complement to the high quality voice mobile service. SMS functionality, linked to voice messaging systems, opens up a new vista of service opportunities and will be a major feature of service offerings in the future. One current example is the delivery of a voice message waiting signal to the mobile, which is sent when the mobile reactivates into the network, indicating without intrusive interruption, that a message has been left while the mobile was unavailable.

Such a feature - and there are many variations of voice and data messaging that can be exploited - begins to put into customers' hands a telecommunications product over which they can exert control and yet can be given the assurance of being contactable.

1.8   User requirements

Market research reveals a consistent set of features which form a vision of personal communications recognised by the majority as representing their ideal individual communication system. However, any practical technology will favour some attributes at the expense of others. Thus PCN sets out to satisfy the most important aspects from a low cost base.

In addition, where compromises have to be made, solutions which can be improved over time from predictable technology advances, such as better semiconductors, are sought. For example, it was considered more acceptable for early handsets to be slightly larger, or provide less than the desired battery life, rather than for the network never to be able to offer contiguous coverage.

PCN is essentially a single service replacing the mobile, cordless and fixed phone by an individual pocket or handportable phone. The emphasis is very much on the hand portable not the mobile car phone, although phones are expected to work if the user is in a vehicle and car adapters are an option.

The network provides high quality speech, conceptually as good as that of the fixed PSTN. This requires good quality voice coding with little degradation from the mobile network implementation. Thus PCN radio coverage must be very good with its service area covered contiguously with an adequate radio signal strength. The radio coverage supports handsets both outdoors and in buildings. The geographic coverage is targeted on built up environs and regularly visited areas such that there is a simple marketable proposition to the customer that the phone will work anywhere within a clearly specified area.

PCN aims to create a mass market for mobile phones. Any restrictions or complications in the ways in which the phones can be used will limit its take up. This means that the handset should be able to always make or receive calls anywhere within its service area, with its basic method of use identical to that of using a conventional fixed phone. In addition, calls once established should be maintained whether the user is stationary or mobile. Other features are provided as enhancements to this basic service.

These can include alerting lights or tones associated with cheap tariff zones, call waiting etc. A particularly attractive service is an integral voicemail within the network. This service provides two important features:

- There are times when calls to a personal phone can be an intrusion, e.g. during important meetings, whilst at the theatre, etc. By diverting the call to a voice-mail system and using an alerting light on the handset, the message can be accessed - when convenient

- If the PCN phone is outside its geographic coverage area, incoming calls can still be completed by the voicemail system. On returning to the radio coverage area, a message waiting light is activated on the handset, and the voicemail box can be accessed

The handset is the customer's direct contact with PCN, and its attractiveness has a major effect on the potential purchase of the service. Size, weight and ergonomic appearance must all be considered. The physical length of any telephone is governed by the distance between the mouth and ear, although 'flip-phones' can fold to a shorter length for carrying. However, the overall weight and volume are determined principally by the size of the battery. With today's technology for digital phones, the batteries required for a full day's usage, which consists of intermittently making and receiving calls and continuously remaining affiliated to the network, are larger than ideal. As the power consumption of semiconductors continues to fall with succeeding generations of sub-micron feature sizes, handportables will reach the point where further reductions in size and weight will be of diminishing value.

1.9 Pricing

The benefits of a fully mobile pocket phone are clear, but for widespread acceptance the cost of usage must be low. The key is to put in place a low cost infrastructure, using economies of scale to aid its implementation, and high levels of usage to amortise fixed costs over many calls. The tariffing of PCN must be suitable for a mass market. Although a premium is possible for its mobility advantages over the fixed phone, when used near the user's home, the increment over PSTN rates should be small while local and trunk rates should be supported. This requires there to be interconnection between the PCN and the PSTN at sufficient points to access the different tariff bands. Interconnect rates have to be agreed between the PCN and PSTN operators such that they can cover their respective costs for their parts of delivering calls across the two networks. The costs of interconnection to the PSTN are negotiated to be similar to fixed telephone rates, and use local and trunk bands. In contrast, mobile cellular has treated all calls as long distance trunk calls, with high interconnect rates, which lead to high tariffs for this service.

Figure 1.3 shows the options for setting the price of personal communications services. The relative costs of the handset, the monthly rental and call charges can be varied. It has become common for handsets to be heavily discounted to provide a low cost of entry. However, to protect the high dealer commissions paid and the operator's subsidy, the customer must enter a long service agreement or be subject to a handset lock for which an unlocking fee is charged, if the handset is to be used on a competing network.

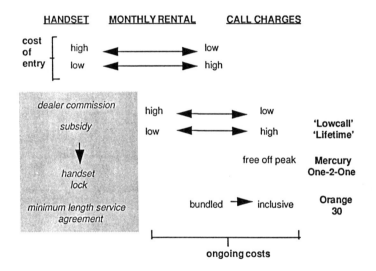

Figure 1.3 Pricing options for personal communications

Churn has been a key parameter in the profitability of mobile operators and has risen above 20%. As personal communications develops as a mass consumer market purchase, it is important to establish propositions encouraging customer loyalty.

PCN companies have concentrated more on the ongoing costs. In the case of Mercury One-2-One the monthly rental and call charges are low and include free off peak local calls. Orange have bundled tariffs with a specified number of call minutes included. Both PCN companies have lower rates for incoming calls than the 900 MHz operators. These tariffs encourage high usage and competition with fixed as well as mobile phones.

As well as services for their traditional customers, the established analogue networks of the 900 MHz operators offer low usage services typically for emergency communication only. Such customers buy subsidised handsets and pay quite low monthly rentals but very high call charges. As they use the network so little they provide a useful incremental revenue to the operators.

Although the PCN operators use digital cellular radio standards, the majority of customers of the established cellular radio operators are still served by analogue networks. Their new GSM networks offer digital quality speech and pan-European and international roaming, but for these operators, continuing to exploit and develop their established analogue networks is extremely profitable. The 900 MHz radio spectrum can be migrated from analogue to digital use but as analogue handsets continue to be sold aggressively, a full transition seems many years away.

1.10   Service positioning

Although new technology and improvements to networks are important factors the dramatic growth in personal communications in the UK has been achieved predominantly by service positioning. The rapid take-up of the service offered by the 1800 MHz digital networks of the two PCN operators has been matched by unprecedented growth of customer numbers on the existing 900 MHz analogue cellular networks. Establishment of 900 MHz digital GSM as a premium service for pan-European roaming has been slower but is used by many of the most profitable customers of cellular networks.

Figure 1.4 considers positioning and the new service packaging options which are driving personal communications take-up. Traditionally, different forms of telecommunication systems were used by the business and consumer sectors when at or away from their normal base locations.

Initially, cellular radio offered a 'business' tariff which appealed to the highly mobile businessperson. The low usage/emergency tariffs have met a clear market niche for the consumer who wants the security and availability of a personal phone for emergency and unforeseen situations. These service packages have had much to do with the growth in customers on the UK 900 MHz analogue networks.

The 1800 MHz PCN operators have a capacity advantage on their networks due both to deploying the latest digital technology and having access to a wider frequency band. This makes high usage service packages most appropriate. By setting low tariffs and relying on high volume low margin usage to achieve profitability, tariffs such as 'Personal Call' have crossed the traditional market segmentation. Consumers gain the advantages of mobility away from home, but due to the low usage costs, substitute PSTN fixed calls with their personal handsets. In addition, businessmen subscribe to this tariff, which although not the lowest cost for day time use, is attractive overall as an all-day, single handset service.

A combined fixed-mobile service is seen by many as a package which will be increasingly sought particularly for business. Routeing of calls to a single personal number on either the fixed or mobile network as appropriate is already possible, but a future service should use a single cellular/cordless handset, switch networks automatically, and select the least cost routeing.

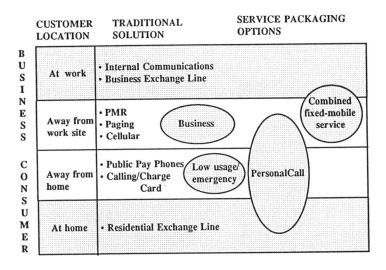

Figure 1.4 Personal communications service positioning

1.11 Personal communication standards

As well as cellular radio based systems, personal communication services may be provided by cordless telephones, pagers, private mobile radio, and satellite links. Cellular techniques are dominant because such networks most fully meet the requirements of the mass market giving the benefits of mobility particularly throughout urban environments. However:

- cordless techniques are the most cost effective solution in limited coverage high density indoor environments

- satellite PCN is the most cost effective solution for wide area, low population density environments

- paging gives highly reliable one way communication over wide areas, and has a long battery life

–   fixed networks can be enhanced towards Universal Personal Telecommunications (UPT) which in its simplest form will enable users to make and receive calls to a personal number at any fixed access point.

All of these technologies influence the path to third generation systems. For example, if higher bit rates are required, new systems may incorporate the benefits of cordless access through very small cells.

1.11.1   Cordless telephones

In the UK, telepoint networks were unsuccessful. The concept of being able to use a cordless telephone away from home suffered from the high cost of complementary home base stations, non-contiguous coverage by telepoint base stations and the inability to receive as well as to make calls. Where some of these shortcomings have been eliminated, telepoint has been more successful. In the French 'Bi-Pop' telepoint system, subscribers can manually register at specific base stations and receive incoming calls. As two-way calling needs additional signalling in the PSTN, this results in additional complexity and cost.

In Japan, the Personal Handy-phone System (PHS) is expected to be successful because the coverage will be extensive and contiguous (several thousand cells of 100-200m radii) while the handsets are very small and light with a long battery life (5 hours talk time, 150 hours standby time). The system is constructed by making the most of the existing public telephone network. Adapter points connect to the PSTN local switches and handle the movement of users by registration, paging, location and authentication activities.

1.11.2   UMTS

Today's personal/mobile communications operators in Europe are licensed to use bands around 1.8 GHz and 900 MHz with some limited allocations at lower frequencies. The standards employed include the digital ETSI specified GSM, DCS1800, CT-2 and DECT, various analogue cellular systems, and the fixed PSTN including UPT features. In addition, higher frequencies have been assigned for wireless LANs and wireless access to the PSTN.

Figure 1.5 shows a European standards road map. The first generation standards were diverse and based on analogue technology. Second generation systems are based on digital techniques in the broadest sense, i.e. digital modulation, time domain based (e.g. TDMA, TDD), digital speech coding, etc. It also coincided with the formation of ETSI such that a single body produced harmonised standards for Europe. The third generation is seen as being a move towards unified systems, offering a full service capability from one system. It has been argued that DCS1800, although just a variant of GSM, has moved beyond second generation as it concentrates on servicing handsets designed for a wider market than just traditional mobile cellular radio.

The standards for the Future Public Land Mobile Telecommunications Systems (FPLMTS) and Universal Mobile Telecommunication System (UMTS) are being progressed by the ITU and ETSI, respectively. These standards aim to provide an enabling capability to meet the requirements of third generation personal/mobile telecommunications within the frequency range 1.88-2.2 GHz, i.e. low cost, high capacity, personal mobility services.

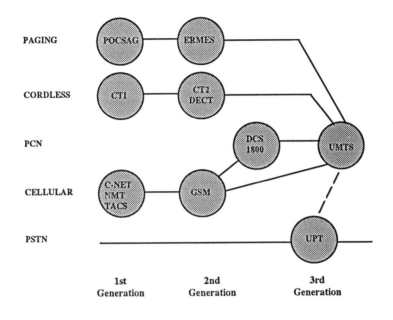

Figure 1.5 Personal/mobile standards

1.11.3 Advanced personal communications

It is generally accepted that the PCN vision is the provision of affordable communications with total freedom and mobility, ubiquitously available and provided in a manner that puts the users in control of their communications. DCS 1800 provided an initial standard to deliver this vision and, for example, enabled Mercury One-2-One to launch the world's first personal communications network in London during 1993.

Third generation systems should continue the move towards systems fully meeting the requirements of personal communications by offering services and features which are demanded by the market, and can be offered economically and competitively. Developers of standards for third generation systems have opportunities to provide technical solutions to respond to the emerging and projected market needs, but will also need to take into account the significant investments that many will have made in digital infrastructure for personal, mobile and fixed network services. Existing standards work and network infrastructure will continue to be exploited, where this is appropriate, particularly bearing in mind the expected large populations of personal communication terminals that will be in use at the time of the introduction of the next generation of systems.

Acknowledgements

I would like to thank my colleagues at Mercury One-2-One for their valuable contributions toward the successful implementation of PCN and help in preparing this chapter. I also thank the parent companies of Mercury One-2-One, Cable & Wireless, and US West for permission to publish this work.

References

1   European Telecommunications Standards Institute, GSM Technical Specifications, ETSI, 06921 Sophia Antipolis, Cedex, France

2   Ramsdale, P.A. 'Personal Communications in the UK - Implementation of PCN using DCS1800', *Int J. Wireless Information Networks*, vol 1, no 1, January 1994, pp 24-36

3   Ramsdale, P.A. and Harrold, W.B. 'Techniques for Cellular Networks Incorporating Microcells', IEE Conf. PIMR 92, Boston, October 1992

Chapter 2

Coverage prediction methodologies

David E A Britland and Raymond C V Macario

Introduction

The computation of the pattern of coverage of a radio transmitter, be it the base station or the mobile station, is fundamental to the planning of all mobile radio services. Several attributes are looked for, namely:

- actual coverage of the radio signal above a specified signal level
- presence of this signal in regions where it would constitute co-channel interference
- ability of multi-sited stations to fill in completely a specific geographical region
- the possibility of quasi-synchronous operation of certain services.

There are now a wealth of computational methods for this type of activity, whether one is planning a new TV service or a microcellular installation. In fact, if all was straightforward, a single universal algorithm would suffice, but this is not the case. Various development strategies are being deployed, many of which are not generally referenced in the public domain.

The purpose of this chapter is not of course to divulge any particular method, but rather to review the fundamental principles which are behind all such attempts to give a good prediction of radio coverage requirements.

Because in all instances the calculation of signal level from a transmitter, of a given power P_T, antenna gain G_T, at a distance d, is very complicated, when one goes into the matter with thoroughness, only fast running and efficient computer algorithms stand any chance of success of providing area, or radio cell, coverage dimensions. Also certain well established physical principles are always present and this allows one to begin with a very simple model of the earth, and then note the modifications required as the ground becomes hilly or cluttered, for example.

2.1 Wave propagation over the ground

The magnitude of an electromagnetic field radiated from a radiating source placed near the surface of the earth depends on the power of the transmitter, its polarisation, the operating frequency, the heights of both the transmitter and receiver antennas, the range, the electrical characteristics of the earth, the geographical profile, and the gradient of the refractive index in the troposphere. The latter comes from the humid atmosphere close to the ground, of which the first few hundred metres only is generally of interest.

The problem is usually simplified by dividing the path taken by the wave along the earth into three zones, taking into consideration the relevant factors affecting the field variation in each region, shown in Figure 2.1.

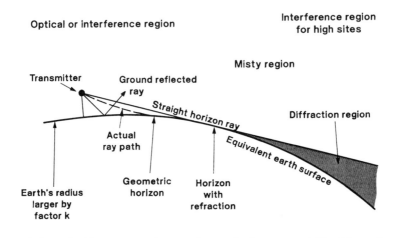

Figure 2.1   Radio wave path over the equivalent curved Earth's surface, the radio horizons, and the propagation zones

This section now outlines the mathematical treatment of the problem if one considers a smooth earth with a regular troposphere. Some well-known semi-empirical methods then follow, since the ground is rarely smooth. To complete the matter, a direction for a more sophisticated solution is described.

2.1.1 The interference region

For distances well within the optical region (Figure 2.1), the field at a distance d can be found by ray theory. The field is considered to be the sum of two parts: that due to a direct ray, and that due to a ray reflected from the ground.

Figure 2.2 shows that this view can apply to a slightly undulating terrain, for example. Therefore, this region is also called the interference region. The field

strength (volts per metre) may be expressed relative to the free space field, which
is generally recognised by the equation

$$E_0 = \frac{\sqrt{30P_T}}{d} \quad V/m \tag{2.1}$$

where d is in metres and P_T is the effective radiated power in watts.

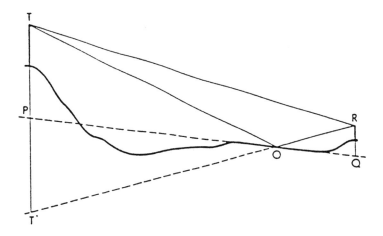

*Figure 2.2   Graphical determination of the point of reflection and two rays in
the interference region for well sited transmitter and receiver*

By introducing an amplitude factor F , called the propagation factor

$$F = \frac{E_r}{E_0} = \left| 1 + Dr\,e^{-iy} \right| \tag{2.2}$$

where ψ is the angle of phase change due to ray path difference, ρ is the reflection
coefficient of the ground reflected ray, and D is a factor that accounts for the
divergence of the rays due to the spherical shape of the earth [1, 2, for example],
the resultant field can be calculated.

It has to be noted that the reflection coefficient ρ is a function of the electrical
characteristics of the earth, and depends upon the wave polarisation. If we assume
a small grazing angle, typically less than 2°, then $\rho \approx -1$, irrespective of the
polarisation, and for short distances it can also be assumed that $D \approx 1$; then
Eq. (2.2) can be written as:

$$F = 2 \left| \sin\left(\frac{2\pi h_1 h_2}{\lambda d}\right) \right|^2 \tag{2.3}$$

where h_j, $j = 1,2$, is the height of the transmitting and receiving antennas. λ is the wavelength and d is the distance, all having the same units. This equation shows that the magnitude of the field strength might oscillate until:

$$d = \frac{4h_1h_2}{\lambda} \tag{2.4}$$

Eq. (2.3) also leads to the well-established fourth power law of propagation loss in the interference region [3], which is independent of the frequency. This is to some extent a deficiency of the ray theory, and also it cannot be extended to treat regions beyond the radio horizon, i.e. the diffraction region. It is, however, a very popular first order approximation, and also shows how antenna height helps propagation coverage.

2.1.2 The diffraction region[1]

In the diffraction region, the field is given by summing an infinite number of modes [4]. Each term includes three factors: a distance factor, which depends on the range between the two antennas, and two height gain functions to account for the heights of the transmitting and receiving antennas. To help the discussion, let us introduce the following normalisation factors x_0 and z_0 for the range and the heights, respectively:

$$x_0 = \left(\frac{a_e^2 \lambda}{\pi^2}\right)^{1/3} \tag{2.5}$$

$$z_0 = \frac{1}{2}\left(\frac{a_e \lambda^2}{\pi^2}\right)^{1/3} \tag{2.6}$$

where a_e is the effective earth radius in metres.

Now the general solution to find the propagation factor F due a vertical dipole over a smooth conducting earth can be expressed in terms of the normalised heights z_1 and z_2, and normalised range x, as:

$$F(x,z_1,z_2) = 2\sqrt{\frac{\pi}{x}} \left| \sum_{n=1}^{\infty} f_n(z_1)f_n(z_2)\exp(\tfrac{1}{2}(\sqrt{3}+i)a_nx) \right| \tag{2.7}$$

$z_{1,2}$ and x are given by: $z_{1,2} = \dfrac{h_{1,2}}{z_0}$, $x = \dfrac{d}{x_0}$,   respectively.

---

[1]We are indebted to Mr Mohammed Bataineh, Yarmouk University, for assistance with this section.

a_n, appearing in the exponential factor in Eq. (2.7), is the nth zero of the Airy function $Ai(u)$, defined as:

$$Ai(u) = \frac{1}{\pi} \int_0^\infty \cos(\frac{1}{3}t^3 + ut)\,dt \qquad (2.8)$$

and the functions $f_n(u)$ are given by:

$$f_n(u) = \frac{Ai(-a_n + e^{i\pi/3}u)}{e^{i\pi/3} Ai'(-a_n)} \qquad (2.9)$$

For distances well beyond the optical range, the terms of the series decrease rapidly and all except the first can be neglected. Therefore Eq. (2.7) can be written as:

$$F(x,z_1,z_2) = 2\sqrt{\frac{\pi}{x}} \left| f_1(z_1)f_1(z_2)\exp(\frac{1}{2}(\sqrt{3} + i)a_1 x) \right| \qquad (2.10)$$

Eq. (2.10) illustrates indeed that the propagation factor consists of three terms: a distance factor which can be easily evaluated and calculated, but two height gain functions involving Airy integrals.

The first zero of the Airy function a_1 is equal to -2.33811 [5]. Then the distance factor can be written in the following form:

$$V(x) = 11 + 10\log x - 17.6x \qquad (2.11)$$

This is exactly the same result which will be found in References [6, 7, 8]. The difficulty of having closed form formulae for the two remaining factors is generally overcome by preparing a set of graphs to find the effect of raising the antennas above the ground, such as may be found in [1] and [9]. To be able to use these graphs in a computer program, empirical equations are fitted to the graphs. Ekstrom [6] used the height gain factor given in [9] and noted the following equation for the case $z < 0.6$:

$$f_1(z) = z \cdot \sqrt{1 + \frac{1}{(hl)^2} + \frac{2}{hl}\sin(\theta)} \qquad (2.12)$$

where $h = h_1$ or h_2, and

$$l = \sqrt{\frac{2\pi p'}{\lambda}}$$

$$p' = \frac{2\pi}{\lambda} \frac{[(\varepsilon_r - 1)^2 + (60\sigma\lambda)^2]^{1/2}}{[\varepsilon_r^2 + (60\sigma\lambda)^2]^b}$$

b = zero for horizontal polarisation (HP), and unity for vertical polarisation (VP), whilst θ is particularly dependent, i.e.

$$\theta_{HP} = \frac{\pi}{4} - \frac{1}{2}\tan^{-1}\frac{\varepsilon_r - 1}{60\sigma\lambda}$$

$$\theta_{VP} = \frac{5\pi}{4} + \frac{1}{2}\tan^{-1}\frac{\varepsilon_r - 1}{60\sigma\lambda} - \tan^{-1}\frac{\varepsilon_r}{60\sigma\lambda}$$

ε_r = the earth's dielectric constant relative to unity (free space)
σ = earth conductivity in mhos/m

For other values of z, the equations given in [10] may be used, with:

$$\begin{aligned} f_1(z) &= 4.3 + 51.04[\log z / 0.6]^{1.4} & 0.6 < z < 1 \\ &= 19.85(z^{0.47} - 0.9) & z \geq 1 \end{aligned} \tag{2.13}$$

There then remains a region between the two above named regions, sometimes termed the intermediate or misty region, for which no direct solution is available. The procedure to predict the field strength in this region has often been achieved by extending the plot from the region in which interference methods are valid, through the intermediate region, into the region of validity of the diffraction methods by bold interpolation [1, 8]. However, for frequencies used in cellular radio fortunately this region hardly exists.

2.1.3 Practical observation

It is possible to observe the above predictions in very flat and calm regions of the world, such as exist in some Middle-East areas. For example, Figure 2.3 shows the observed signal strength from a transmitter tower of 34 m height observed up to a distance of 30 km, well beyond the radio line-of-sight [11]. On a logarithmic basis, the first order terrestrial path loss formula in dB, namely, the fourth power propagation law, is:

$$L_{dB} = 40 \log d_m - 20 \log h_1 h_2 \tag{2.14}$$

The slope of the signal graph γ (using Lee's [12] nomenclature) should be 40 dB/dec, as indeed it is. The slope rolls over to closer to 80 dB/dec as the signal enters the diffraction region. The ground is unlikely to be perfectly smooth, neither is the troposphere always normal, i.e., the earth's atmospheric correction factor $k=4/3$ [1, 2] may not apply over the path.

The above analysis applies in the case of propagation above a smooth spherical earth. Under other conditions, semi-empirical methods have been proposed to calculate the loss of the radiated power during its journey from the transmitter to the receiver. Before describing some of these models, the effect of obstacles in the transmission path is considered.

2.2　Propagation over terrain obstacles

Hills and other obstacles along a radio wave propagation path will cause reduction in field strength, the more so at UHF [13]. Frequently the amount of clearance (or obstruction) is described in terms of Fresnel zone clearance ratio [14]. The Fresnel zone clearance ratio is the ratio of the distance from the direct line between antennas to the path obstacle (distance h in Figure 2.4) and the Fresnel radius r [15]. It is expressed in terms of the parameter v where

$$v = \sqrt{2}\,\frac{h}{r} \tag{2.15}$$

and r is the radius of the nth Fresnel zone, given by

$$r = \sqrt{\frac{n\lambda d_1 d_2}{d_1 + d_2}} \tag{2.16}$$

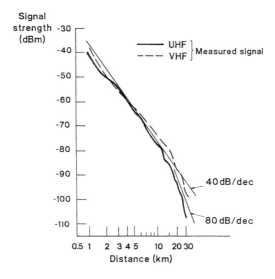

Figure 2.3　　*Observed uniform reduction in signal strength with distance over a very flat desert path*

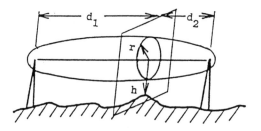

Figure 2.4   Fresnel zone about the line connecting the two antennas

If a straight sharp edge screen is interposed between the transmitter and the receiver, the received field is usually given in terms of an integral of the form:

$$F = \frac{1+i}{2} \int_{v}^{\infty} \exp(-i\frac{\pi}{2}t^2)$$ (2.17)

The approximate values of the diffraction loss for different ranges can be computed from the following equation [16]:

$$
\begin{aligned}
F &= 0 & v \geq 1 \\
&= 20\log(0.5 + 0.62v) & -1 \leq v \leq 0 \\
&= 20\log(0.4 - \sqrt{0.1184 - (0.1v + 0.38)^2} & -2.4 \leq v \leq -1 \\
&= 20\log(0.225 / v) & v \leq -2.4
\end{aligned}
$$

This factor can be taken into account at each point as one moves further away from the transmitting site.

2.3   Empirical propagation models

Empirical methods have been worked out to find the effect of various environments on the propagation of waves. Each of these methods is applicable, but has constraints and suffers from limitations. In what follows, some of these models are presented.

2.3.1 Egli model

Egli introduced the concept of the average irregular terrain [17], and derived a semi empirical formula for the path loss, of the form:

$$L = 88 + 40\log d + 20\log f - 20\log h_1 h_2 \qquad (2.18)$$

This model provides an estimate of the median loss between ground based antennas over gently rolling terrain, with average hill height of approximately 15 m. Note the similarity to the basic formula, Eq. (2.14).

2.3.2 Okumura model

After extensive measurements in and around Tokyo, Okumura [18] proposed the following equation for the path loss:

$$L = A_m(f,d) - H_b(h_b,d) - H_m(h_m,f) \qquad (2.19)$$

where A_m is the median attenuation relative to free space in an urban area over quasi smooth terrain with a base antenna height of 200 m (quite high) and a mobile antenna height 3 m. H_b and H_m are the height gain factors for the base and mobile antenna heights, h_b and h_m, respectively.

All these factors are given in graphs. For different earth profiles Okumura suggested adjustment factors for Eq. (2.19), also given in graphs. The only drawback in Okumura's model is that it was not designed to be used on a computer, but this difficulty was overcome by Hata [19], who put the model to computational use.

2.3.3 Hata model

The analytic approximation for the path loss of Okumura's model has been given by Hata [19] as a set of approximations. The propagation loss in an urban area is used as a standard formula, given by:

$$L = 69.55 + 26.16\log f - 13.82\log h_b - a(h_m) + (44.9 - 6.55\log h_b)\log d$$
$$(2.20)$$

In this equation $a(h_m)$ is the correction factor for h_m. For mobile antenna heights in a small city $a(h_m)$ is given by:

$$a(h_m) = (1.1\log f - 0.7)h_m - (1.56\log f - 0.8)$$

while in a large city, where the building height average is more than 15 m, $a(h_m)$ is given by:

$$a(h_m) = 8.29\,(\log 1.54\,h_m)^2 - 1.1 \qquad\qquad f \le 200\,\text{MHz}$$

$$= 3.2(\log 11.75 h_m)^2 - 4.94 \qquad\qquad f \ge 400\,\text{MHz}$$

In suburban and open areas, the propagation loss is given by:

Suburban: $L = L(\text{urban area}) - 2(\log f / 28)^2 - 5.4$

Open area: $L = L(\text{urban area}) - 4.78(\log f)^2 + 18.33 \log f - 40.94$

Note the signal slope factor $\gamma = 40$, when $h_b = 5.9\,\text{m}$. This model was later adopted as the so-called standard CCIR model [3].

2.3.4 Ibrahim and Parsons model

Based on measurements undertaken in some British cities, two models were produced [20]. The first (empirical) gives the median path loss between isotropic antennas as:

$$L = 47.7 - 8 \log h_b - 20 \log f \cdot h_b + [40 + 14.15 \log(\frac{f+100}{156})] \log d$$
$$+ 0.265L - 0.37H + k \qquad\qquad (2.21)$$

where $k = 0.087U - 5.5$ for the highly urbanised centre of the city, otherwise $k=0$. L is the land usage factor, varying between 0 and 80%. U is the degree of urbanisation, and H is the difference in height between the base and the mobile.

A second (semi-empirical) gives the median path loss as:

$$L = \text{theoretical plane earth loss} + \beta \qquad\qquad (2.22)$$

where $\beta = 20 + f / 40 + 0.18L - 0.34H + k$

Here $k = 0.094U - 5.9$ for the highly urbanised city centre, otherwise $k = 0$.

2.3.5 Bullington model

This is an early theoretical model based on a smooth earth propagation theory, but which includes the approximations for estimating the effects of hills and other obstructions in the radio path [13]. The interesting feature is that this method avoids multiple hills by constructing a single equivalent knife edge [14]; tropospheric effects are accounted for by means of graphs. The prediction procedure was initially given as nomograms.

2.3.6 Longley-Rice model

This model is based on well established propagation theory with atmospheric and terrain effects allowed for by a very large data bank of empirical adjustments [21]. The propagation path is divided into three regions in this model, like Figure 2.1, namely, line-of-sight, diffraction, and forward scatter region, and the field strength is predicted by applying different linear formula path loss in each region, and according to type of location.

2.3.7 Lee model

This method, like Okumura's, is based on measurements in specific environments. The measurements have been undertaken in North American cities [12]. Lee gives a general path-loss equation based on the signal power P_r received relative to the power received at 1 mile intercept P_{r0}:

$$P_r = P_{r0} - g\log(r/r_0) - n\log(f/f_0) + a \qquad (2.23)$$

γ is the path loss slope factor described above, r is the distance from the site, f_0 is the reference test frequency, equal to 900 MHz, r_0 the 1 mile intercept distance, and α is the correction factor to account for height-gains of the base and mobile antennas and different transmitter power. One notes the concept is empirical, because P_{r0} is put in from measurement, in a sense. As is well known, the path loss slope changes from city to city, and one environment to another [15, 16]; therefore the above formula has to be applied in a step-by-step manner along any particular path, e.g. Figure 2.5. Within a propagation program these steps come up according to chosen environment, but use data based on previous measurements, i.e. Table 2.1.

Table 2.1 Programmed environmental conditions according to Lee's model

ENVIRONMENT	1-MILE INTERCEPT	PATH-LOSS SLOPE
Free space	-45.0 dBm	20.0 dB/dec
Open area	-49.0 dBm	43.5 dB/dec
Suburban area	-61.7 dBm	38.4 dB/dec
Urban (Philadelphia)	-70.0 dBm	36.8 dB/dec
Urban (Newark)	-64.0 dBm	43.1 dB/dec
Urban (Tokyo)	-84.0 dBm	30.5 dB/dec

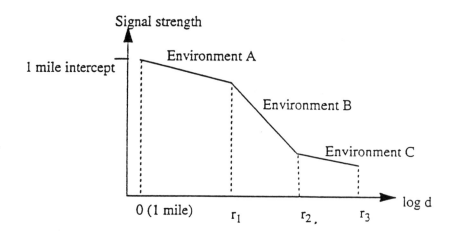

Figure 2.5   Received signal strength due to path loss against log distance

2.4   Topological data banks

The ability of geographic information centres to produce very detailed ordnance survey data, in computer readable form, is well known. We do not need to discuss this in any detail here, except to say that there is an awful lot of data involved, and this can make the production of a propagation coverage diagram a slow process, unless advanced computer practices are employed.

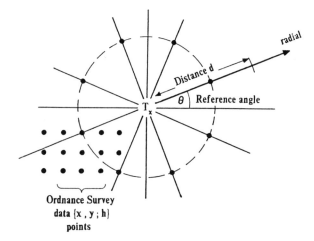

*Figure 2.6   Radial lines for planning cell coverage overlaid on topographic
data points*

For example, Figure 2.6 shows some conceptual ordnance survey data points, which could be every 10 metres, say, and in each case a propagation path is formed by one radial. Using 1° radials clearly requires calculating 360 paths.

To illustrate operation, we now turn to the coloured diagrams with this chapter. Thus Figure 2.7(a) shows the contour path along a particular radial, which actually looks much more rugged on the screen than it would be standing on the path (the vertical scale is in metres, whereas the horizontal is in kilometres). Forestation can be added (green lines) point by point as it occurs. The computer can then produce, on command, the attractive relief maps, an example of which is shown in Figure 2.7(b). Superimposed is the predicted radio coverage from a particular hill top site. This type of technology is now, of course, well established.

2.4.1 Options available

Figure 2.8(a) shows a screen shot on which the various propagation models described above can be selected by pointing, and a multi-site display can be set up, i.e. in the example, at 424 MHz. The interesting feature, which of course is not seen in the next photographs, is that there is actually not too much difference for the propagation model predictions, at least in the area which these displays cover.

Figure 2.8(b) shows the effects of coverage reduction due to co-channel interference. In the left hand picture both radio sites are transmitting on the same carrier frequency and the resultant network coverage is limited. In the right hand picture the two sites have been assigned different frequencies and the area served is greatly increased. Although in both cases the radio frequency field strengths are the same in the area between the sites, in the left hand case, the signals are unusable due to the co-channel interference encountered.

On the other hand, in Figure 2.9(a) the coverage of the two cells, over a fairly flat terrain, is almost circular, and matters like adjacent channel protection and quasi-synchronous co-channel operation can be investigated. In quasi-synchronous or simulcast systems, three reception areas are characterised within the RF coverage.

(i)   The capture zone, shown in yellow, where the difference in signal level between transmitters is sufficiently large that one swamps the other and only one is heard. This is particularly true for FM transmission which exhibits a 'capture effect' in the receiver discriminator.

(ii)   The non-capture zone, shown in green, where the signals from the two transmitters are received at similar levels and substantially at the same time. In an analogue system the received audio exhibits excessive flutter, as if there is considerable multipath, and causes some distortion. In digital systems, if suitable equalisers and coding are present, the data may be decoded correctly.

(iii)   The distortion zone shown as a small red area indicates where the signals are received at similar levels, as above, but also suffer from time differences due to their transmission path lengths. These are sufficient to cause intersymbol interference in a digital network and severe audio distortion in an analogue one.

Figure 2.9(b) shows how by means of three sites, operating at different frequencies, greater coverage and network capacity may be achieved. Coverage from each transmitter is indicated by a different raster pattern and signal field strength by colour. This is known as 'best server plot' which is useful in cellular networks for estimating regions where 'handovers' are likely to occur.

However, if the terrain has more 'aggressive' features, differences are noticeable. Thus Figure 2.10 puts side-by-side the coverage predicted from a site in the centre of the city of Hull. Going SW from the centre is a pair of ridges, i.e., Figure 2.11, and a distinctive difference is apparent between the Bullington method (Section 2.3.5), and the Okumura method (Section 2.3.2), because the former, as described, makes them equivalent to one ridge.

It is important to note that the predicted signal coverage around a base station does not fall off regularly, as for example in Figure 2.3. The signal is very scattered in general, even in the primary coverage areas (yellow regions shown in the plots reproduced here). Also the prediction methodologies outlined here refer in the main to macrocells. Microcellular operation is discussed in Chapter 10.

One matter, which is important even in macrocells, is building penetration loss [22]. This is a complex matter, but some discussion on the subject will be found in Chapter 5. Indeed, as said there, 'coverage is perceived as one of the most important measures for comparing the merit of competing networks', and this is why an assured prediction, i.e. not too optimistic, nor too pessimistic, is rather important. In general, because of building penetration loss and the personal carrying of mobiles in cars or overcoats, say, this needs a signal level allowance built in any coverage prediction.

Very detailed terrain-based propagation models are also reported, based on an integral equation approach e.g. [23, 24], or finite difference scheme e.g. [25]. These methods require even more intensive computing, as discussed below.

2.5 Effect of the troposphere

The above models also do not account for the effects of the troposphere on the signal path. The troposphere has an influence because of the variation of its refractive index with height. Under normal propagation conditions the refractive index influence may be included by modifying the earth radius [1, 2]. Since, as is well-known, the refractive index is a function of basically three meteorological parameters, namely, humidity, pressure and temperature, which may at times deviate from their normal variation, and lead to a non-standard refractive index profile, i.e. propagation under anomalous conditions.

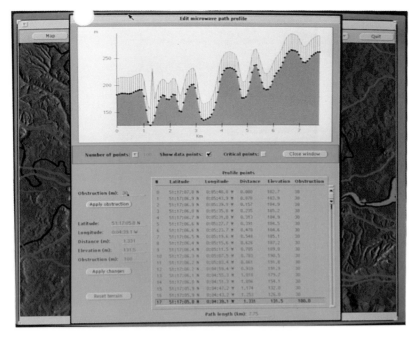

Figure 2.7(a) The ground contour for a particular path, with forestation added

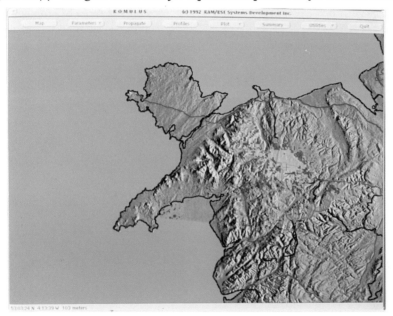

*Figure 2.7(b) A relief map constructed from topological data, and the radio
coverage predicted for the site marked (+)*

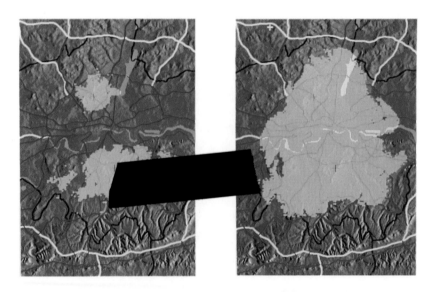

Figure 2.8(a)   A propagation prediction screen, showing the menu of propagation

Figure 2.8(b)   Screen displays to show the effective reduction in coverage due to co-channel interference

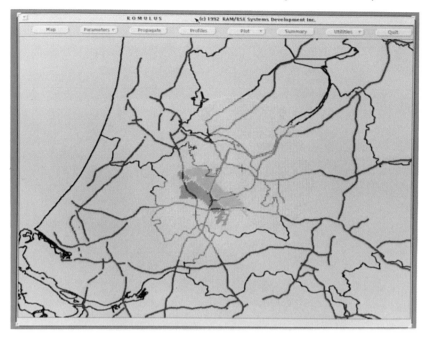

Figure 2.9(a) A simulcast coverage situation (over a flat landscape) using two transmitters at 440 MHz

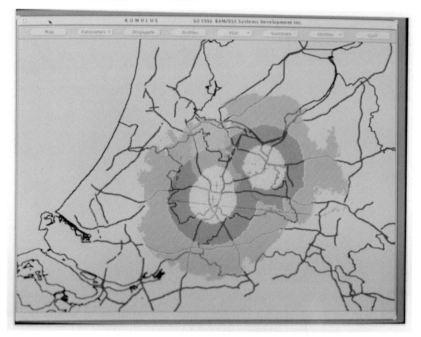

Figure 2.9(b) A best server plot showing three transmitter sites

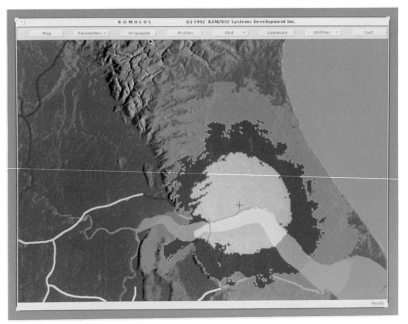

*Figure 2.10(a)   Prediction for a 200 MHz radio cell operating in the city of Hull,
using the Bullington procedure*

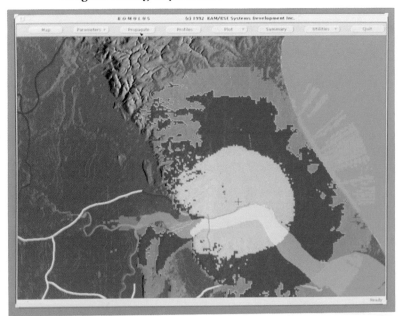

*Figure 2.10(b)   Prediction for a 200 MHz radio cell operating in the city of Hull
using the Okumura procedure*

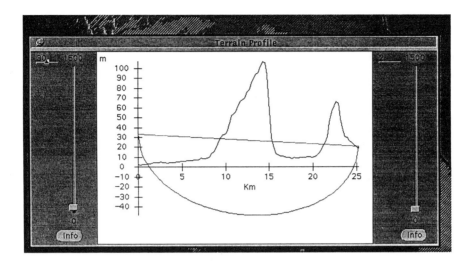

Figure 2.11   *The ground profile going SW from the transmitter site in Figure 2.10*

Ducting is the most dramatic example of propagation of radio waves in non-standard or anomalous conditions, and results in a significant variation in the received field intensity [26].

Consideration of refractive index variability with propagation prediction is now possible by using a full wave method, known as the parabolic equation.

2.5.1 The parabolic equation

A full wave solution to the field strength applicable to all the regions of space, taking into consideration the variation in refractivity, is given by the parabolic equation method. A historical development of the method may be found in [27], and a detailed derivation of it is given in [28]. Parabolic equation calculations can include the effects of the transmitter/receiver geometry, the frequency, polarisation, antenna patterns, earth's curvature, and surface roughness [29]. Computer codes have been developed based on the mathematics involved, especially by Craig and Levy [30, 31]. Results obtained using this method, compared with earlier methods, will be superior, but do require far greater computational effort, and fortunately are perhaps not so necessary in temperate latitudes, but could well be of growing importance in warmer climates [1, 11, 21].

2.6 Summary

The chapter has attempted to provide a summary of the essential ingredients of mobile radio coverage. Only a selection from an enormous list of possible references is given.

The chapter, hopefully, has also indicated the extent of this subject, and also at the opportunity for further fundamental studies and measurements.

References

1   Kerr, D. *Propagation of short radio waves*, McGraw-Hill, (1951)

2   Reed, H. and Russell, C. *Ultra high frequency propagation*, John Wiley and Sons, (1962)

3   Parsons , J. D. *The mobile radio propagation channel*, Pentech Press, (1992)

4   Van der Pol, B. and Bremmer, H. 'The diffraction of electromagnetic waves from an electric point source round a finitely conducting sphere with application to radio telegraphy and the theory of the rainbow, Part 1', *Phil. Mag.*, vol. 24, pp.141-176, (1937)

5   Fock, V.A. *Electromagnetic diffraction and propagation problems*, Pergamon Press, (1965)

6   Ekstrom, J.L. 'VHF-UHF propagation performance predictions for low altitude communication links operating over water', Milcom'93, Boston USA, pp.605-608, (1993)

7   Domb, C. and Pryce, M. 'The calculation of field strength over a spherical earth', *J. IEE*, vol. 94, part III, no. 31, pp.325-336, (1944)

8   ITU draft revision of recommendation ITU-R PN. 526-2, 'Propagation by diffraction', ITU document 5/61-E, (1993)

9   Norton, K. 'The calculation of ground-wave field intensity over a finitely conducting spherical earth', *Proc. IRE*, vol. 29, pp. 623-640, (1941)

10   Blake, L. V. *Radar range-performance analysis,* Artech House, (1986)

11 Macario, R.C.V. 'Trying to bridge the Gulf with GSM', *Mobile Middle East and Africa Mag.*, vol. 1. no. 4, pp. 17-20, (1994)

12 Lee, W.C.Y. *Mobile communications design fundamentals*, Howard W. Sams & Co., (1986)

13 Bullington, K. 'Radio propagation for vehicular communications', *IEEE Trans. Veh. Tech.*, vol. 26, pp. 295-308, (1977)

14 Parsons, J.D. Chapter 1 in R. J. Holbeche (Ed.), *Land mobile radio systems*, Peter Peregrinus Ltd., (1985)

15 *IEEE Vehicular Technology Society*, 'Special issue on mobile radio propagation', vol. 37, February, (1988).

16 Lee, W.C.Y. *Mobile communications engineering*, McGraw-Hill, (1982)

17 Egli, J. 'Radio propagation above 40 Mc over irregular terrain', *Proc. IRE*, vol. 45, pp. 1383-1391, (1957)

18 Okumura, Y. Ohmori, E. Kawano, T. and Fukuda, K. 'Field strength and its variability in VHF and UHF land mobile service', *Rev. Elec. Comm. Lab* 16, pp. 825-873, (1968)

19 Hata, M. 'Empirical formula for propagation loss in land mobile radio services', *IEEE Trans. Veh. Tech.*, VT-29, pp. 317-325, (1980)

20 Ibrahim, M. and Parsons, J.D. 'Signal strength prediction in built-up areas', *Proc. IEE, Part F*, vol. 130, no. 5, pp. 377-385, (1983)

21 Longley, A. G. and Rice, P. L. 'Prediction of tropospheric radio transmission loss over irregular terrain- a computer method', *Inst. Telecom. Sci., Essa Tech. Rep.* ERL79-ITS67, Boulder, Co, (1968)

22 Rice, L.P. 'Radio transmission into buildings at 35 and 150 Mc', *The Bell System Technical Journal*, vol. 38, pp. 197-210, (1959)

23 Hviid, J.T., Andersen, J.B., Toftgrad, J. and Bojer, J. 'Terrain-based propagation model for rural area - an integral equation approach', *IEEE Trans.* AP-43, no. 1, pp. 41-45, (1995)

24 Moroney, D. and Cullen, P.J. 'An integral equation approach to UHF coverage estimation', ICAP, Publication no. 407, pp. 367-372, (1995)

25   Marcuse, S. W. 'A hybrid (finite-difference-surface Green's function) method for computing transmission losses in an inhomogeneous atmosphere over irregular terrain', *IEEE Trans*. AP-40, no. 12, pp. 1451-1458, (1992)

26   Hall, M.P.M. *Effect of the troposphere on radio communications*, Peter Peregrinus Ltd., (1979)

27   Tappert, F. 'The parabolic approximation method', in *Wave propagation and underwater acoustics*, Lecture Notes in Physics, Springer Verlag, pp. 224-287, (1977)

28   Kuttler, J.R. and Dockery, G.D. 'Theoretical description of the parabolic equation approximation/Fourier split-step method of representing electromagnetic propagation in the troposphere', *Radio Science*, pp. 381-393, (1991)

29   Levy, M. 'Horizontal parabolic equation solution of radiowave propagation problems on large domains', *IEEE Trans*., AP-43, no. 3, pp.137-144, (1995)

30   Craig, K. 'Propagation modelling in the troposphere: Parabolic equation method', *Electronics Letters*, vol. 24, pp. 1136-11391, (1988)

31   Craig, K. and Levy, M. 'Parabolic equation modelling of the effects of multipath and ducting on radar systems', *Proc. IEE, Part F*, vol. 138, no.2, p. 153-162, (1991)

Chapter 3

Modulation and multipath countermeasures

Alister G Burr

3.1 The role of modulation

Modulation and coding occupy a fundamental place in any mobile or personal communication system. They mediate between the data services, which are to be provided, and the radio systems which form the means of that provision. It is the modulation and coding which determine the quality of the data services, and also how they interact with the radio channel. In particular they determine the resource requirements of the system, in terms of both RF power required and bandwidth occupied.

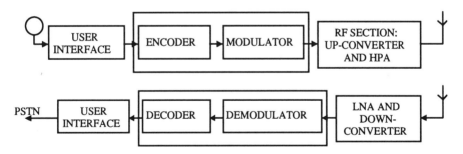

Figure 3.1   Block diagram of a communication system, showing encoder/modulator and decoder/demodulator

Figure 3.1 shows a radio communication system in block diagram form, indicating the position of the modulator/encoder and the demodulator/decoder. Traditionally, coding and modulation have been regarded as two quite separate functions, performed by separate subsystems. However, since the introduction of combined coding and modulation, of which more later, at the beginning of the last decade [37], it has become accepted that they should properly be considered one entity; hence the boxes shown within the dotted areas.

One may regard the encoder/modulator as a single sub-system, which maps data presented to it, by means of a user interface, onto a modulated RF carrier for subsequent processing, amplification and transmission, by the RF sub-system. The demodulator/decoder conversely takes the received RF signal and performs the inverse mapping, back to a data stream for onward transmission. This view underlines the importance of the process; it is the encoding and modulation that determines, for example, the bandwidth occupied by the transmitted signal, whilst it is the demodulator/decoder which determines the quality of the resulting data service availability in terms of BER and delay. It also determines the robustness of the system to channel impairments, due both to the RF sub-systems (such as phase noise and non-linearity) and the RF channel (such as multipath dispersion and fading). Thus the correct choice for the modulation/coding scheme is vital to the efficient operation of the whole system.

3.2   Mobile radio channels

The object of the modulation/coding system is to match the transmitted signal to the characteristics of the radio channel. We must, therefore, consider the characteristics of the mobile radio channel, and in particular, the degradations that may arise; in fact, the mobile radio channel is one of the most difficult encountered by a radio system [24], in terms of the range and severity of degradations encountered. This is because the need for mobility and the locations in which the system is used mean that a line-of-sight radio path is rarely available. Also there are the requirements for minimal resource utilisation; i.e. low-cost, low power mobiles and good spectral efficiency [14, 15].

Noise is of course encountered in any communication system including thermal noise from the RF sub-system in the receiver, if from no other sources. In a cellular mobile/personal communication system, however, the limiting factor is usually not thermal noise, but *co-channel interference* (CCI), from surrounding cells [30]. The robustness of the receiver to CCI is one of the main factors determining the capacity of such a system. In addition there may be interference from adjacent channels within the same cell (ACI).

The need for power efficiency in the mobile dictates the use of power-efficient high power amplifiers (HPAs) in the mobile, which generally suffer from non-linearity. Hence modulation and coding schemes which are not affected by this are highly desirable. This also requires power, as well as bandwidth-efficient, modulation and coding to be used.

However, the most significant problem in mobile radio systems is due to the channel itself. This is the problem of *multipath*, caused by the existence of multiple paths between transmitter and receiver, which may be subject to different time delays and phase shifts [30]. This gives rise to two deleterious effects: *fading* and *dispersion.* Fading is due to the interference of the signals from two or more paths, which can cause various degrees of cancellation. Dispersion is due to the relative time delay on some paths, which may cause *intersymbol interference.* This can cause nulls in the channel frequency response, hence called *frequency-selective fading.* A large part of this chapter is devoted to these effects and their influence on the various modulation and coding schemes. In the next section the behaviour of the channel is considered in more detail.

Note that the influence of multipath depends on the relationship of the symbol rate to the channel. If the system bandwidth is less than the bandwidth of any frequency selective fades, the system is described as *narrowband.* In this case the fading effects are more significant than the dispersion. Conversely, a *wideband* system is affected by both dispersion and fading. This distinction will be used later in the chapter.

3.3   Multipath channels

In personal and mobile radio systems, indeed, it is rare for there to exist one strong line-of-sight path between transmitter and receiver. Usually several significant signals are received by reflection and scattering from buildings, etc. And then there are *multiple paths* from transmitter to receiver [30]. The signals on these paths are subject to different delays, phase shifts and Doppler shifts, and arrive at the receiver in random phase relation to one another. The interference between these signals gives rise to a number of deleterious effects, one of the most serious problems of the mobile radio channel, known collectively as *multipath.* The most important of these are *fading* and *dispersion.*

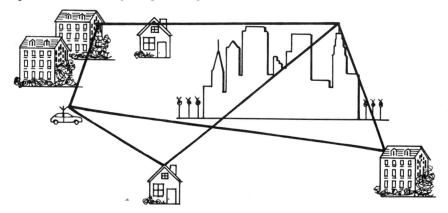

Figure 3.2 Origin of multipath

Fading is due to the interference of multiple signals with random relative phase. Constructive and destructive interference between them causes random variations in the amplitude of the received signal. This will increase the error rate in digital systems, since errors will occur when the signal-to-noise ratio drops below a certain threshold. Dispersion is due to differences in the delay of the various paths, which disperses transmitted pulses in time. If the variation of the delay is comparable with the symbol period, delayed signals from an earlier symbol may interfere with the next symbol, causing *intersymbol interference* (ISI).

If the transmitter, or receiver, is in motion (which of course is most likely in the case of a mobile radio system), then the relative phase shifts of the different paths will change with time, potentially quite rapidly. The different components of the signal are thus subject to different Doppler frequency shifts, because of the differences in their angle of arrival [10]. Thus overall the signal is subject to a Doppler spread, such that a single transmitted frequency is received as a band of frequencies. This makes carrier recovery difficult in coherent systems, and may also give rise to errors in non-coherent demodulators [24].

3.3.1   Multipath channel modelling

It is helpful to create mathematical models of the multipath channel, for three main purposes. The first is to aid in the understanding of the channel, and its effects on communication signals. Secondly, it allows one to analyse these effects, and derive results mathematically. Thirdly, the models may form the basis for computer simulation of the channel, which can be used in situations which are too complex to analyse mathematically.

There is a hierarchy of available models, starting from the most detailed and general, and descending to the most specific and simplest [40]. The relationship between them is shown in Table 3.1.

Table 3.1 Mathematical models of the multipath channel

$$\textit{Linear time - variant system} \qquad h(\tau,t)$$

$$\downarrow v_{max}\tau_{max} \ll 1$$

$$\textit{Quasi - stationary} \qquad h_t(\tau)$$

$$\downarrow \text{uncorrelated scatterers}$$

$$\textit{GWSSUS} \qquad \sum_i h_i(t)\delta(\tau - \tau_i)$$

$$\downarrow \text{Sampled}$$

$$\textit{Tapped delay line} \qquad \sum_{j=0}^{N-1} a_j(t)\delta(\tau - jT)$$

$$\downarrow W_S\tau_{max} \ll 1$$

$$\textit{Narrowband model} \qquad h_0(t)$$

The most general models the channel as a *linear time-variant system* [4]. This is described by its *time-variant impulse response*, $h(\tau, t)$. This gives the response of the channel at time t, to an impulse at time $t - \tau$, and therefore gives the channel impulse response, and shows how it varies with time. $h(\tau, t)$ is also related to three other functions, which give the same information, but in a different form. These are illustrated in Table 3.2, where F denotes Fourier transformation with respect to the subscripted variable. The most interesting of these functions are $H(f, t)$, the *time-variant frequency response*, and $S(\tau, v)$, the *scattering function*. Since v can be interpreted as Doppler shift, the latter function gives the Doppler spectrum of the received signal as a function of the delay τ.

Table 3.2 Relationship of functions describing channel

$$
\begin{array}{ccc}
h(\tau,t) & \xrightarrow{\;F_\tau\;} & H(f,t) \\
F_t \downarrow & & F_t \downarrow \\
S(\tau,v) & \xrightarrow{\;F_\tau\;} & B(f,v)
\end{array}
$$

If the time-variation of the channel is slow, it may be treated as *quasi-stationary*, or *piece-wise stationary*; in other words, as a linear system whose parameters vary with time, but which are constant for periods of a few transmitted symbols. One can strengthen this condition, because if the maximum Doppler shift, v_{max}, is much less than the inverse of the maximum delay in the channel, τ_{max}, then the channel is said to be *separable*, so that the delay parameter τ and the time t can be treated separately. If we further assume that the signals on the different paths are uncorrelated, and have Gaussian distributions, we have the well-known Gaussian wide-sense stationary uncorrelated scatterers (GWSSUS) model [4, 12]. Here the impulse response may be treated as the sum of a series of impulses with delays τ_i, representing the different paths, with amplitude/phases h_i, which vary with time, giving the form shown in Table 3.1. This is illustrated in Fig. 3.3.

The squared amplitude of this impulse response is known as the *power-delay profile* of the channel. Various models have been used for the power-delay profile in particular environments. The best-known for mobile radio channels are the COST 207 profiles, used in the definition of GSM (hilly terrain, typical urban, bad urban, etc.) [11]. More arbitrary shapes, such as exponential, Gaussian or two-ray have also been used. For indoor channels, statistical models have been developed in which the arrival of rays is treated as a Poisson process.

The most important parameter of the power-delay profile response is the *delay-spread* Δ, which is the standard deviation of the delay [6]:

$$
\Delta = \sqrt{\dfrac{\int_0^\infty (\tau - D)^2 \left| h(\tau,t) \right|^2 \, d\tau}{\int_0^\infty \left| h(\tau,t) \right|^2 \, d\tau}} = \sqrt{\dfrac{\sum_i (\tau_i - D)^2 \left| h_i \right|^2}{\sum_i \left| h_i \right|^2}}
\tag{3.1}
$$

where D is the mean delay. The inverse of this is the *coherence bandwidth*, because it gives the frequency range over which the frequency response of the channel is likely to remain flat.

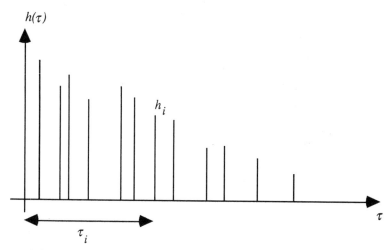

$h(\tau)$

h_i

τ

τ_i

Figure 3.3 Impulse response for GWSSUS model

For most purposes the response of the channel outside the signal bandwidth of the signal is of little importance. Thus one can sample the channel at the symbol rate, leading to the *tapped delay-line model* [4, 35]. Here the delayed signals on the different paths are lumped into delays of multiples of the symbol period. This can be represented by a tapped delay line (Figure 3.4) with delays of one symbol period. Each tap is weighted by a (complex) coefficient a_j, $j = 1 .. n\text{-}1$, and multiplied by a *fading envelope*, which implements the time-variable element of the channel. For the GWSSUS channel, this is a complex Gaussian process. Its spectrum is given by the Doppler spectrum at the given delay; i.e. by the scattering function $S(\tau, v)$. Various models have been used for this [25, 35], but the bandwidth is of the order of the maximum expected Doppler shift [40].

The coefficients a_j are derived from the GWSSUS coefficients according to:

$$a_j = \sqrt{\sum_{i=0}^{\infty} g^2(jT - \tau_i)\overline{|h_i|^2}} \qquad (3.2)$$

where $g(t)$ is the shape of the signalling pulse at the receiver. This model, as well as forming the basis for analysis, is readily implemented as a computer simulation of the channel.

The final simplification of this model occurs when the symbol period T is much greater than the channel delay spread, or equivalently, when the signal bandwidth is much less than the coherence bandwidth. Under these conditions dispersion is negligible, and only one tap is required in the model, and hence the tapped delay-line reduces to the first branch only. This may be called the *narrowband model*. Note that for quite a wide range of channels, those with delay spreads comparable with the symbol period, a two-tap model is appropriate

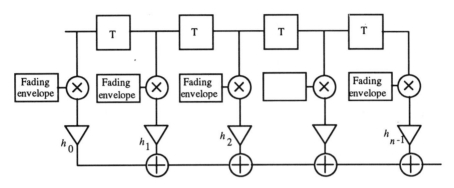

Figure 3.4 Tapped delay-line model

These models highlight the differences between narrowband and wideband signals in terms of how they are affected by a multipath channel, as mentioned above. They may, however, be defined more precisely. 'Narrowband' means that the bandwidth is much less than the coherence bandwidth of the channel, whilst 'wideband' means that it is comparable with, or greater than, the coherence bandwidth. Narrowband signals are not subject to *dispersion*, only to *fading*; wideband channels are subject to both dispersion and fading. Equivalently, narrowband channels have a frequency response flat across the signal bandwidth; wideband channels do not: they are described as *frequency-selective*. This may be an advantage for very wideband systems, where the delay-line model has a large number of taps, since it implies that not all the spectrum is likely to fade simultaneously. Some types of equaliser for wideband systems, such as the RAKE receiver in CDMA systems, can utilise this effect to mitigate the effects of fading. These issues are dealt with below for the two types of channel.

3.3.2 Fading

We discuss first the effects of multipath on narrowband signalling, namely fading, and the main technique for its mitigation, namely *diversity*. Fading arises because the signals on the multiple paths interfere constructively and destructively at random. Since there are a very large number of paths, which are normally assumed to be independent, these signals add to give a random process with a complex Gaussian distribution. The amplitude r of the received signal, therefore, has the Rayleigh distribution [31] which is the distribution of the amplitude of a complex Gaussian process given by

$$p(r) = \frac{r}{\sigma^2} \exp\left(-\frac{r^2}{2\sigma^2}\right) \tag{3.3}$$

where σ^2 is the total power in the multipath signal. If, however, there is a significant line-of-sight signal, not subject to multipath, the signal amplitude will have the Rician distribution [31, 34]:

$$p(r) = \frac{r}{\sigma^2} \exp\left(-\frac{r^2 + s^2}{2\sigma^2}\right) I_0\left(\frac{rs}{\sigma^2}\right) \tag{3.4}$$

where s^2 is the power of the line-of-sight component, and I_0 denotes the zeroth order modified Bessel function of the first kind.

The spectrum of this process depends on the variation of the channel with time. Equivalently, it can be regarded as the Doppler spectrum of the received signal. Since, in general, it comes from a different direction relative to the motion of the mobile, each path will be subject to a different Doppler shift (Figure 3.5) [10]. Hence the received signal suffers a Doppler 'spread', rather than a single Doppler shift. As mentioned above, this spectrum is given by the scattering function $S(\tau,\nu)$, substituting in the value of τ appropriate to the delay.

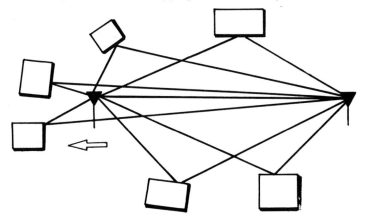

Figure 3.5 Doppler shift of components

The effect of fading is firstly to increase the bit error ratio (BER), because during fades the signal drops below the threshold signal-to-noise ratio required to maintain a low BER. It can be shown [31] that for coherent PSK on a Rayleigh channel the BER P_e is given by:

$$P_e = \frac{1}{2}\left(1 - \sqrt{\frac{E_b/N_0}{1 + E_b/N_0}}\right) \tag{3.5}$$

This is plotted against bit energy to noise density ratio as curve (a) below in Figure 3.20. Note, however, errors on a fading channel tend to occur in bursts, lasting for the duration of a link fade.

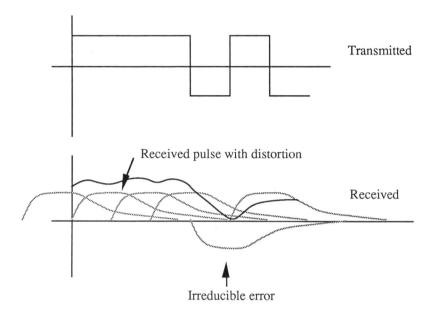

Figure 3.6 Intersymbol interference and irreducible errors due to dispersion

3.3.3 Dispersion

Wideband channels were defined above as channels in which the signal bandwidth is comparable to, or greater than, the coherence bandwidth of the channel. Thus the channel is frequency-selective. Equivalently, it means that the symbol period is comparable to, or less than, the delay spread Δ. Hence the signal is subject to dispersion, as well as to fading. The appropriate model is the tapped delay line of Figure 3.4. The dispersion gives rise to *intersymbol interference* (Figure 3.6). The transmitted rectangular pulses are distorted by the channel, and received with a long 'tail' due to multipath dispersion. This interferes with the reception of subsequent symbols, and may give rise to errors (termed *irreducible errors* because they cannot be eliminated by increasing the signal power - see below). It has been shown that for most mobile radio systems the irreducible BER is proportional to the square of the ratio of the delay spread to the symbol period (Figure 3.7) [6]. This assumes Rayleigh fading, but is valid over a wide range of symbol rates regardless of the actual shape of the power-delay profile.

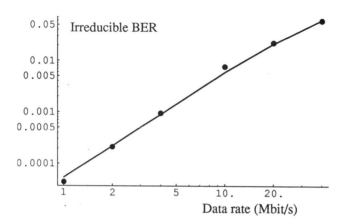

Figure 3.7   *Irreducible BER for QPSK with no Nyquist filtering and a channel*
delay spread of 20 ns

3.4   Modulation techniques

3.4.1   Parameters

We next consider the parameters of a modulation/coding scheme by which its
effectiveness may be judged.   These can then be used to provide a set of
specifications for the scheme and the sub-systems that implement it.

Perhaps the most important is the bandwidth requirement of the scheme, since
a deficiency in this respect cannot be overcome anywhere else in the system.   The
bandwidth requirement is determined by the *spectrum* of the modulated signal,
usually presented as a plot of Power Spectral Density (PSD) against frequency,
Figure 3.8.

Ideally, of course, the PSD should be zero outside the band occupied.   In
practice, however, this can never be so, and the spectrum extends to infinity.   This
is either because of the inherent characteristics of the modulation scheme (see
below), or because of the practical implementation of filters (which must have a
finite roll-off rate).   Hence we must define the bandwidth such that the signal
power falling outside the band is below a specified threshold (as shown).   In
practice, this threshold is determined by the tolerance of the system (and any others
sharing the band) to adjacent channel interference (ACI), which is itself a feature of
the modulation/coding scheme.

The other main parameter of a modulation/coding scheme is the *bit error ratio*
performance. BER is defined as the ratio of erroneous bits received, to the total
number of bits received.   It is also known more loosely as the 'error rate'.   It is
equal to the probability of bit error, P_b.   Frequently it is plotted as a logarithmic

plot, against signal-to-noise ratio (SNR), leading to the well-known 'waterfall curve' (Figure 3.9). In fact, the ordinate of the graph is normally the *bit energy to noise density ratio* E_b/N_0, since this results in a more system-independent measure.

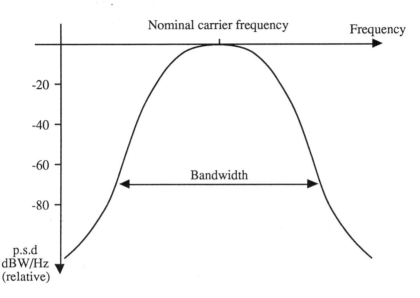

Figure 3.8   Spectrum of modulated signal

The shape of these curves depends on the channel and on the modulation scheme. Curve (a) in Figure 3.9 is the curve typical for the Average White Gaussian Noise (AWGN) channel. Curve (b) is the shape that often occurs on fading channels [31], while curve (c) shows an 'error rate floor', where the BER tends to a finite limit as the bit energy to noise density ratio increases. This can occur, for example, on channels subject to dispersion. The level of the 'floor' is called the *irreducible BER*.

Ideally one would like the BER of the service to be zero, but this does not occur in practice. Hence one must specify a *required* BER. This in general will be different for different services. For example, digital speech services can usually tolerate a BER of 10^{-3}, or higher, while for data transfer the standard may be 10^{-9} (to be equivalent to fixed services), or 10^{-6} in some systems [23]. Then for the given required BER, the curve can be used to obtain the required bit energy to noise density ratio, which influences the link budget calculation.

Note that the 'noise' here may not only be thermal noise. In a cellular system co-channel interference from users in other cells is in fact much more important. In this case the required bit energy to noise density ratio determines the reuse distance of the cellular system, and hence the cluster size.

Other characteristics of a modulation/coding scheme are also of particular importance in mobile radio systems [8]. Firstly, the high power amplifiers (HPAs) used in mobile handsets are usually highly non-linear, because of the requirement for power efficiency. These amplifiers give rise to AM-AM and AM-PM conversion, which may result in an irreducible BER floor. The optimum solution is to use a constant envelope modulation scheme, which does not give rise to these effects. This means that phase-only modulation should be used. However, some schemes are used which are not truly constant envelope, but which have been designed to minimise envelope variations. Note that the requirement for true constant envelope modulation inherently gives rise to a spectrum of unlimited bandwidth [31].

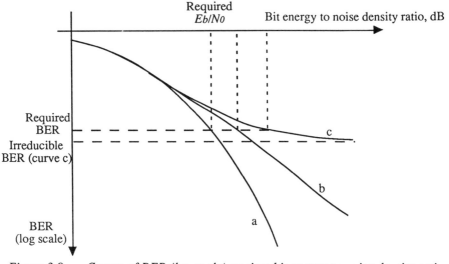

Figure 3.9   *Curves of BER (log scale) against bit energy to noise density ratio (a) white Gaussian noise channel; (b) fading channel; (c) showing irreducible BER floor*

3.4.2 *Linear and exponential schemes*

Modulation schemes can be divided into two classes, often termed *linear* and *exponential* [8, 31]. The terms describe the mathematical relationship between the data and the modulated signal, but the distinction is perhaps most easily understood in terms of amplitude/phase modulation versus frequency modulation. Linear modulation is equivalent to amplitude/phase modulation. Each data symbol to be transmitted is represented by a particular state of amplitude and phase of the transmitted carrier. These states are conveniently represented as points on a *constellation diagram* (Figure 3.11). There the radius from the origin gives the carrier amplitude, while the angle from the positive horizontal axis, marked with an arrow, gives the phase.

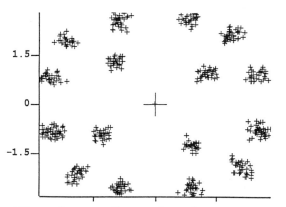

Figure 3.10   Effect of a non-linear channel on a non-constant envelope (16QAM) constellation

The constellation diagram may be regarded as an Argand diagram, in which the amplitude/phase states are represented as complex numbers (the horizontal axis the real part, the vertical the imaginary).   The term *linear* then comes from the mathematical representation of the transmitted carrier as the product of this complex number d_i, which is determined by the data, with the carrier signal waveform:

$$a(t) = d_i \cos(\omega_c t) \qquad (3.6)$$

where ω_c is the carrier angular frequency (radians/s).

If the same state of amplitude/phase is maintained for a full symbol period, then the scheme can be regarded as modulation by a series of rectangular pulses. This gives rise to a $\sin x/x$ shaped spectrum, Figure 3.12(a), with large sidelobes. Hence the signal is usually filtered to minimise its bandwidth. Nyquist filters are mainly used to avoid the ISI.

Figure 3.12(b) and (c) shows two forms of Nyquist filtered pulse. They result in a smooth transition between phase/amplitude states, in the time domain, which prevents linear schemes from maintaining constant envelope, even if all the symbols have the same amplitude (as in the $\pi/4$ QPSK schemes).

Note that, where there are $M > 2$ points in the constellation, i.e. M different symbols may be transmitted, then $\log_2(M) > 1$ bit is transmitted, per symbol period. Schemes with more than 4 constellation points are generally called *multilevel schemes*.

An exponential modulation scheme is equivalent to frequency modulation. The instantaneous frequency is varied by a small *deviation frequency f_d*, either side of the nominal carrier frequency f_c. Such schemes are variants of *continuous-phase frequency-shift keying* (CP-FSK), in which there are no phase discontinuities during symbol transitions. (The term 'exponential' arises because the frequency shift may be modelled mathematically by multiplying the carrier by $\exp(2\pi j d_i f_d t)$, where $d_i = \pm 1$, the current data.

We may define a *modulation index h* for CP-FSK schemes. This is defined as the ratio of the total frequency deviation to the data rate r_b:

$$h = \frac{2 f_d}{r_b} = 2 f_d T \qquad (3.7)$$

where T is the bit (symbol) period. The smallest usable value of h is 0.5, and CP-FSK with this index is called *minimum-shift keying* (MSK). MSK has the property that the phase changes by $\pm\pi/4$ during each symbol period.

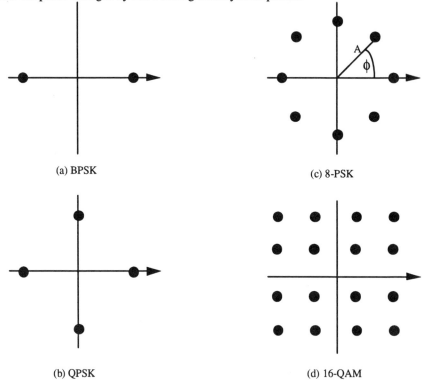

(a) BPSK

(c) 8-PSK

(b) QPSK

(d) 16-QAM

Figure 3.11   Constellation diagrams for linear modulation schemes

The signal can be generated most straightforwardly by applying the digital baseband data signal to a frequency modulator. Because only frequency is varied, and not amplitude, these schemes are clearly constant envelope. However, if the signal is filtered after modulation, to improve its spectrum occupancy, envelope variations are introduced. For this reason, a pre-modulation filter is often applied to the baseband data signal [1]. The resulting family of schemes are generally called *continuous-phase modulation* (CPM), of which the best-known and most important is *Gaussian minimum-shift keying* (GMSK) [28].

The relative advantage of the linear and exponential modulation schemes hinges almost entirely upon the constant envelope property of the exponential plan with the resulting inherent spectral broadening. Linear modulation schemes can be made significantly more bandwidth-efficient by appropriate filtering, but are not constant envelope, and therefore require linear (or linearised) HPAs. Both linear and exponential schemes are currently proposed, or in use, in cellular mobile radio systems, in the form of π/4 QPSK and GMSK, respectively. These are described in more detail below, and the rationale for their use in a mobile radio system is given.

3.5 Coherent and non-coherent demodulation

We consider next demodulation techniques and two basic types [13] are discussed in this section.

Coherent demodulation requires a reference carrier signal in the demodulator. This allows the demodulator to make use of all available information in the received signal, and thus can provide *maximum-likelihood* detection; i.e. an ideal coherent demodulator yields the minimum possible probability of error. Coherent demodulation is illustrated in Figure 3.13(a): the received signal is multiplied by 90° reference signals, which recover the *in-phase* and the *quadrature* components of the signal, from which amplitude and phase may be extracted, and the data recovered.

However, a carrier reference is not usually available, and must be recovered from the received signal itself, which is a difficult procedure, and indeed, the carrier recovery process may be very complex, and usually constitutes the major part of a coherent demodulator. Further, in a mobile radio system the signal, as we saw, is subject to rapid phase shifts due to fading, which the carrier recovery process has to track, which in some cases is impossible, even in principle.

Hence, in practice, mobile radio systems usually use *non-coherent* demodulation. Here no reference is available, and the performance is therefore to some degree sub-optimal, since some information is discarded. However, the loss in performance is generally small, of the order of 1-2 dB, and receiver complexity is substantially reduced. Further, non-coherent schemes are generally much more robust to fast fading, than are coherent demodulators [8, 24].

In most modulation schemes, employed in mobile radio, the information is encoded primarily in the phase of the signal. However, in the absence of a reference carrier, the demodulator cannot make use of absolute phase information, and hence indirect means of extracting the phase information must be used. There are two main techniques:

Time domain pulses (symbol levels -1, 1):

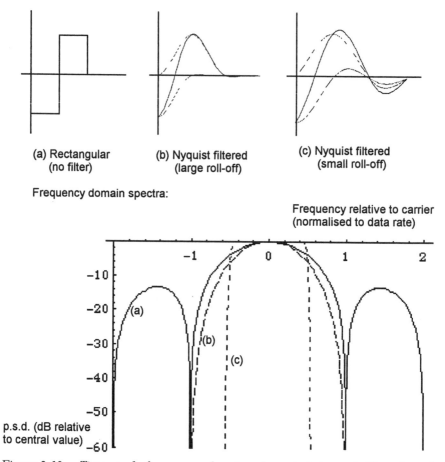

(a) Rectangular (no filter)

(b) Nyquist filtered (large roll-off)

(c) Nyquist filtered (small roll-off)

Frequency domain spectra:

Frequency relative to carrier (normalised to data rate)

p.s.d. (dB relative to central value)

Figure 3.12   *Time and frequency domain characteristics of filtered and unfiltered linear modulation schemes*

Differential demodulation. Here differential phase information is used, i.e. the difference between the phases of the current and the previous symbol. Then assuming that the previous symbol was correctly demodulated, the current symbol may also be demodulated. A block diagram is shown in Figure 3.13(b): it is essentially the same as the coherent demodulator except that the reference is provided, via a symbol-period delay element, by the previous signal.

Limiter-discriminator demodulation. Here frequency information is used, rather than phase, but since frequency is the rate-of-change of phase, this can be used to recover phase information. Hence it can be used for linear schemes where the information is carried by the phase, as well as exponential schemes, such as CP-FSK, where frequency is modulated directly. Essentially the modulation is treated as digital FM, using an FM discriminator, preceded by a limiter since amplitude variations are unwanted (Figure 3.13(c)).

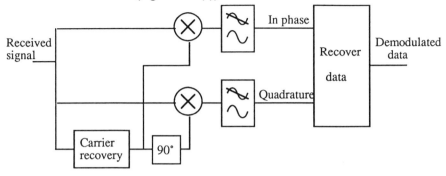

(a) Coherent demodulator

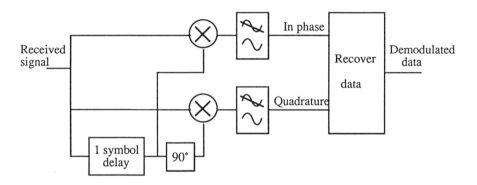

(b) Differential demodulator

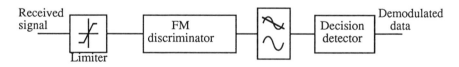

(c) Limiter-discriminator demodulator

Figure 3.13   Demodulator structures: (a) coherent; (b) differential; (c) limiter-discriminator

3.6 Modulation methods

We now consider two most important modulation schemes: those proposed, or in use, in cellular mobile radio systems. These are $\pi/4$ QPSK, used for one of the

North American AMPS-D second-generation standards, and GMSK, used in the
GSM system. The former is linear, the latter exponential; the differences between
the two exemplify the relative advantages of the two types of scheme.

3.6.1   π/4 QPSK

π/4 QPSK is a variant of QPSK, see Figure 3.11(b). However, successive symbols
are relatively shifted in phase by 45° (π/4). The result of this is illustrated in Figure
3.14, which shows the phase-amplitude trajectories between successive symbols in
the constellation diagram. It can be seen that the trajectories never pass through
the origin, which reduces the possible amplitude variation, compared with QPSK.
This should reduce the degradation when non-linear amplifiers are used.

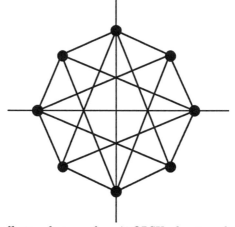

Figure 3.14 Constellation diagram for π/4 QPSK, showing phase trajectories

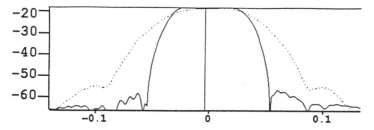

*Figure 3.15   Spectra of π/4 QPSK, roll-off factor 0.35 (solid line) and GMSK,
time-bandwidth product 0.3 (dotted)*

π/4 QPSK is also used with Nyquist filtering, which maintains a narrow
bandwidth. This was in fact the rationale for its choice in the North American
system. It was desired to maintain the same channelisation as the analogue AMPS

system, but achieve a significant increase in capacity. The use of π/4 QPSK allows the 30 kHz AMPS channel to be shared, using TDMA, by three D-AMPS users.

3.6.2 GMSK

GMSK is CP-FSK with $h = 0.5$ (i.e. MSK), in which the FM modulator is preceded by a Gaussian response lowpass filter [28]. This is an exponential modulation scheme, and as such is constant envelope. The Gaussian filter has the effect of minimising the bandwidth required, but being constant envelope, it necessarily has a wider bandwidth than a linear scheme like π/4 QPSK. Figure 3.15 compares the spectrum of both schemes.

A continuous phase modulation scheme, like GMSK, can conveniently be represented by a *phase trellis diagram*. This is essentially the eye diagram of the modulated signal phase, being the ensemble of the phase trajectories for all possible data sequences. Figure 3.16 shows an example of a phase trellis for GMSK. Note that the phase trellis diagram for MSK would be a tessellation of full 'diamond'-shaped eye openings; the effect of the pre-modulation filter is to round off the vertices and reduce the width of the eye openings. This makes it more difficult to demodulate. With a simple demodulator, a degradation of up to about 1 dB is noted, although the full performance may be recovered using a Viterbi maximum-likelihood sequence estimator (MLSE), taking advantage of the similarity of the actual phase trellis to the trellis diagram used in decoding convolutional codes.

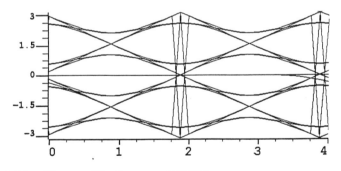

Figure 3.16   Phase trellis diagram of GMSK

3.7 Coding techniques

We now consider the principles of coding, which can be considered as a separate function from modulation but is really best integrated with the modulation. Note also, we are considering *error-control*, or *channel coding,* only, and not speech coding (*source coding*) nor *cryptographic coding* (for security).

There are two basic types of error-control code [36]: *forward error correcting* (FEC) codes, which can correct errors even if there is a forward channel only, and *error-detecting* codes, which detect errors only, and hence require a feedback channel to signal a re-transmission (an automatic repeat request, or ARQ system). The disadvantage of error detecting codes is that the delay inherent in re-transmissions is excessive for speech services; hence we will consider FEC coding only.

3.7.1   Rationale

The basic function of error-control coding, as the name suggests, is to reduce the number of reception errors in a digital communications system.   FEC coding is used in all current cellular systems, particularly so in GSM, to protect those bits of coded speech which are more important, in terms of their effect on the quality of reproduction (known as class I bits) [22].

However, coding has another function, which in fact is more important in most applications.   It is clear that BER may normally be reduced by increasing the signal-to-noise ratio (in accordance with Figure 3.9).   The advantage of coding, however, is that the same BER may be achieved for a lower SNR in a coded system, than in a comparable uncoded system.   This may allow the power budget to be relaxed, giving a number of potential system advantages.   This advantage given by a coded system is measured as *coding gain* [17].

The coding gain of a coding scheme is defined as 'the reduction in bit energy to noise density ratio E_b/N_0 in the coded system compared to the uncoded, for a given BER, for the same data rate'.   Note the use of E_b/N_0 rather than SNR: this ensures that any differences in the bandwidth required are automatically allowed for.   Figure 3.17. gives 'waterfall' curves for a coded and an uncoded system, showing the coding gain.   Note the importance of quoting the BER at which the coding gain is measured.   The coding gain varies very significantly with BER, and above a certain level may even be negative.

In fact, most coding schemes give a lower coding gain at higher (poor) BERs. Since speech services can generally operate adequately with quite high BERs i.e. GSM [22], it has often been perceived that there is little gained by FEC coding in mobile telephony systems.   However, the potential of coding, particularly integrated with modulation, in this application has recently begun to be realised.

3.7.2   Block codes

The earliest types of code, and the simplest to understand, are block codes [26]. Here input data is read into the encoder in blocks of length k bits, and mapped into output *codewords* of length $n > k$.   Only certain length n codewords are allowed, and the allowed codewords are chosen so as to differ in as many places as possible.

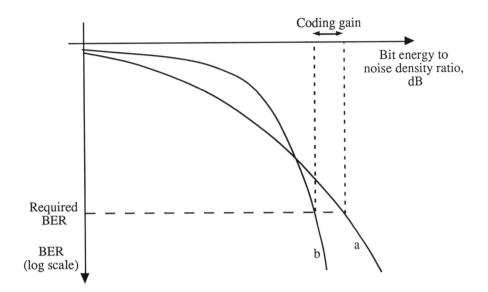

Figure 3.17   Waterfall curves for uncoded (a) and coded systems (b), showing coding gain

For example, suppose $k = 2$ and $n = 5$, and the following 4 codewords are allowed:

Input (Data)	Output (Codewords)
00	00000
01	01110
10	10011
11	11101
	↑

The codewords differ from one another in at least 3 places. Now suppose the word 01110 is transmitted, and one error occurs (in the third position), so that the word 01010 is received. The receiver knows that an error has occurred, because the received word is not one of the allowed codewords. It can also correct it by comparing the received word with each of the codewords. It differs in one place from the correct codeword, but by at least 2 from each of the others. The receiver's 'best guess' will then be the correct codeword, since one error is more likely to occur than two.

The number of places in which two words differ is known as the *Hamming distance* between them (after Richard Hamming, who invented the first practical FEC codes [21]). The minimum Hamming distance between any pair of codewords is known as the *minimum distance d_{min}* of the code. It is clear that a code will correct any number of errors strictly less than $d_{min}/2$.

The above example shows that the introduction of additional bits, or *redundancy*, into the coded data stream is essential. Hence to maintain the same data throughput rate, the coded bit rate must be increased, which will result in a bandwidth expansion. In a bandwidth limited system, such as mobile radio, this is clearly a significant disadvantage. However, more recent techniques such as coded modulation (see next section) can overcome this disadvantage. A measure of the bandwidth expansion introduced is given by the *rate* of the code, defined as the number of coded bits divided by the number of data bits, i.e. $R = k/n$. Rates approaching unity give the least bandwidth expansion.

The most commonly-used block codes are the *Bose-Chaudhuri-Hocquenghem* (universally known as BCH) codes, and the *Reed-Solomon* (RS) codes [26, 29]. These are both based on the algebra of finite fields, and have structures which allow particularly efficient decoding, using shift registers and logic units. RS codes, in fact, treat groups of bits together as *symbols*, and are then able to correct symbol errors. This allows one to correct bursts of bit errors, so long as only a few symbols are affected. Table 3.3 gives the minimum Hamming distance and the coding gain for some BCH codes.

Table 3.3   *Minimum distance and coding gains (for word error rate 10^{-6}) of BCH codes*

n	k	d_{min}	Coding gain (dB)
7	4	3	0.6
15	11	3	1.6
	7	5	1.3
31	26	3	2.0
	16	7	2.6
	11	11	2.6
63	57	3	2.2
	45	7	3.6
	36	11	2.2
127	120	3	2.2
	71	19	4.9
	57	23	4.5
255	247	3	2.2
	189	19	5.5
	131	37	5.5

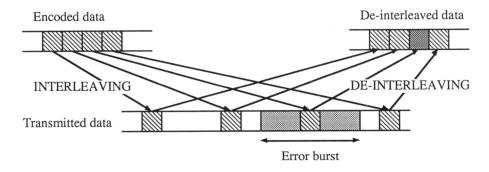

Figure 3.18   Interleaving for burst error correction

Burst error correction in fact is an important feature of codes for the mobile radio channel, since bursts may easily occur due to short-term channel fading. If the code used does not have burst correction capabilities, *interleaving* is often used, whereby the bits of a codeword are distributed throughout the transmitted data stream, and then reconstituted at the receiver. In this way error bursts do not affect an entire codeword, as shown in Figure 3.18.

3.7.3   Convolutional codes

In convolutional codes, in contrast, successive blocks are not treated independently, and so the coded data cannot be split into codewords [26]. Instead the current block of coded data (of length n) depends, not only on the current input block, but also on a number of previous blocks. The encoder consists, at least conceptually, of a shift register surrounded by modulo-2 adders (Figure 3.19). The previous blocks are then stored in the shift register. The total number of blocks on which the output depends, the length of the shift register, is often known as the *constraint length n*.   Note that the input blocks may be of length $k-1$, corresponding to k bits shifted into the register at once.

To replace the concept of codewords, we consider *code sequences*. The object of the decoder is to determine the code sequence closest in Hamming distance to the received sequence, and thus corresponding to the fewest errors. There exists an efficient algorithm, the *Viterbi algorithm* [39, 26], which traces all possible code sequences and returns the closest according to a particular *distance metric*, which could of course be the Hamming distance.

However, the use of Hamming distance as a metric is in fact sub-optimal. In making 'hard decisions' in the demodulator prior to decoding, some information is lost to the decoder on the reliability of the decisions made. If the demodulator passes to the decoder *side information* on reliability, the decoder may be able to make a better estimate of the code sequence. This can be done by means of *soft decision decoding*, which uses *squared Euclidean distance* (SED) as a metric.

SED is a measure of how close the signal is to the actual transmitted signal. The use of soft decision decoding generally increases the coding gain by about 2 dB [9].

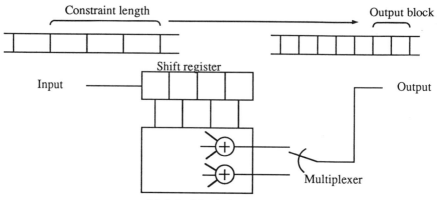

Figure 3.19   Schematic of operation of convolutional encoder

Soft decision can also be used with block codes, but a major advantage of convolutional codes is the ease with which it may be incorporated into Viterbi decoding - only the metric calculation need be changed. For block codes, soft decision decoders are usually several times more complex than hard decision.

3.7.4   Coding in cellular systems

It is interesting that the only application of coding in existing TDMA cellular systems is, as mentioned above, the protection of class I bits of the speech coder. Here the object is to ensure a lower BER for these bits, than for others, but a convolutional code, using hard decision decoding with interleaving to allow for burst errors, is deemed sufficient. The potential advantages in terms of power reductions described above remain largely unexplored, mainly for reasons mentioned above. This is likely to change in any third generation TDMA systems, as will be considered in the final section of this chapter.

However, the situation is quite different in CDMA systems, including the new standard proposed by Qualcomm, and others [19]. CDMA is subject to self-interference from users in the same cell (as well as from outside it). Without coding, the system's tolerance to this interference is poor, which severely limits its capacity. Hence, quite powerful coding is used to increase tolerance to interference, and thus allow many more users. Note also, that in CDMA systems, where the bandwidth is always large, the bandwidth expansion caused by coding is not important. Hence low rate codes (rate $\frac{1}{3}$ or $\frac{1}{4}$) can be used.

3.8 Multipath countermeasures

3.8.1 Diversity

Fading on a mobile radio channel is usually countered by means of *diversity* [25, 30]. The principle of diversity is to provide two or more channels, or *branches*, for the same information signal, subject to statistically independent fading. Then if one path fades, we may expect the other(s) not to, and hence a satisfactory error ratio is maintained. Diversity techniques may be classified according to the means by which the additional channels are provided, and according to how the multiple signals are combined.

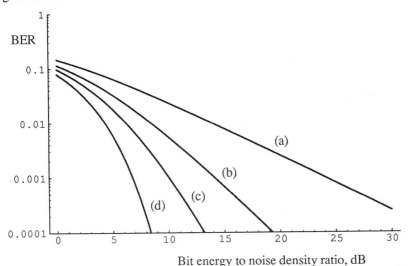

Figure 3.20   *BER versus bit energy to noise density ratio for (a) Rayleigh fading channel, (b) Rayleigh channel with two-branch diversity, (c) with four-branch diversity, compared to (d) AWGN channel*

The three main diversity techniques are called *space, frequency* and *time diversity* (Figure 3.21). The most commonly-used is space diversity [30]. Here, independently-fading signals are provided by receiving the same signal at two or more different antennas, separated in space. Because Rayleigh fading is due to small phase differences in the multipath signals, it can be expected to vary rapidly across space and be almost completely uncorrelated, within the space of a few wavelengths, at the transmission frequency. At the usual frequencies for mobile radio transmission, these separations are feasible at base stations, and sometimes for vehicle-mounted mobiles, but not for handportables.

If space diversity is not feasible, frequency or time diversity may be used. Here the same signal is transmitted on multiple channels, either at different frequencies, or repeated at different times. The frequency or time separation must be sufficient that the fading is uncorrelated; i.e. several times the *coherence bandwidth* $1/\Delta$ in bandwidth, or the *coherence time* in time. The coherence time describes the fading rate of the channel, and is approximately $1/v_{max}$. Frequency and time diversity have the obvious disadvantage that additional spectrum-space is used, which may reduce the overall capacity.

There are also three main combining methods (Figure 3.21) [25], each of which may be used with any of the diversity techniques. The optimum technique, which provides a maximum-likelihood detection method, is *maximum-ratio combining* (MRC). In this scheme the two diversity branches are weighted proportionately to the signal amplitudes, and then added. The weighting minimises the noise added from faded branches. Figure 3.22, curves (b) and (c), show the BER performance of MRC with 2 and 4 branches, respectively. It can be seen that diversity results in plots on the log-log scale, steeper by a factor equal to the number of diversity branches. This results in very significant gains in bit energy to noise density requirements, making diversity highly advantageous, if it can be implemented.

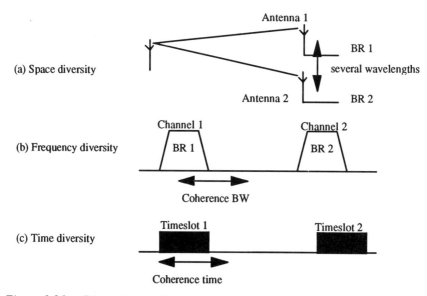

Figure 3.21   Diversity techniques

Clearly MRC requires that the received signal amplitudes are tracked and the weights changed in step. If this is not feasible, *equal-gain* combining (EGC) may be used. Here the branches are equally weighted. Both MRC and EGC require that the signals should be added in phase, or coherently.

The third combining technique, *selection* (or *switched*) *combining* (SC), switches between branches according to which has the strongest signal. This does not require coherence between the branches. Both EGC and SC incur a penalty of a dB or so compared to MRC, but this gives a sizeable advantage over systems without diversity.

The performance of FEC coding in fading channels can also be understood in terms of diversity [31]. If a code has a Hamming distance d_{min}, so that any two codewords differ in at least d_{min} symbols, then any decision in the decoder is taken on the basis of at least this number of symbols. If the symbols fade independently, then this is equivalent to time diversity with d_{min} branches, but with a much smaller penalty in terms of spectrum usage.

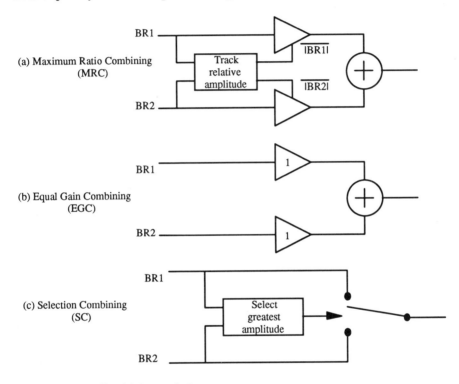

Figure 3.22   Combining techniques

3.8.2 Equalisation

Dispersion can also be countered by correcting the frequency response of the channel. As noted above, a dispersive channel has a non-flat frequency response over the signal bandwidth [20]. This may be corrected by adding a filter in the receiver which exactly cancels the variations in frequency response in the channel, resulting in a flat response overall. This is known as *linear equalisation* (Figure 3.23) [31, 33]. Normally, the equaliser is implemented as an *adaptive finite impulse response* (FIR) filter, with some suitable adaptation algorithm used to select the filter weights.

Unfortunately, the performance of the linear equaliser is very poor on mobile radio channels [7]. The frequency response tends to exhibit deep nulls, which gives rise to compensatory peaks in the equaliser response. These amplify the noise at these frequencies, giving rise to a *noise multiplication* effect. This makes linear equalisers unsuitable for mobile radio channels, except where a strong line-of-sight component exists and can be guaranteed.

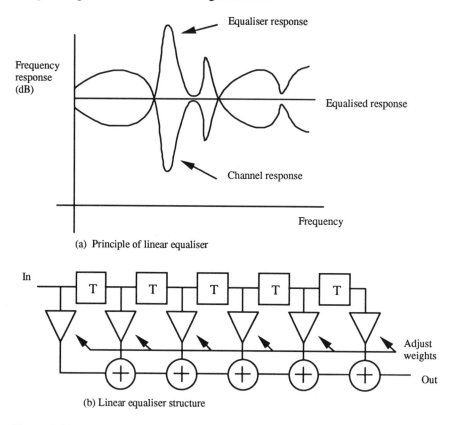

(a) Principle of linear equaliser

(b) Linear equaliser structure

Figure 3.23   Linear equaliser

Thus other techniques have been developed for the mobile radio channel. These are also known as 'equalisers', although strictly they are not, since their principle of operation is quite different. The simplest is the *decision-feedback equaliser* (DFE) (Figure 3.24) [2]. The principle of the DFE is to subtract the intersymbol interference, making use of symbol decisions already made to estimate the ISI. An input matched FIR filter is used to condition the received signal, ISI due to previous symbols is subtracted, and then the current symbol decision is made. The ISI due to this decision is then estimated using another FIR filter, and fed back. The DFE performs much better than the linear equaliser on the mobile channel, but it suffers from *error propagation*, since if an error is made in one symbol, the ISI estimate will be erroneous, and may give rise to further errors.

The optimum equalisation technique is *Viterbi equalisation*, which is *maximum-likelihood sequence estimation* (MLSE) [18], performed by means of the Viterbi algorithm, as used for decoding convolutional codes. The channel may be treated as a finite state machine, in which the different states represent the different combinations of input data. The Viterbi algorithm is a means of tracing all the possible data sequences, and selecting the one which was most likely to have been transmitted (hence the term MLSE).

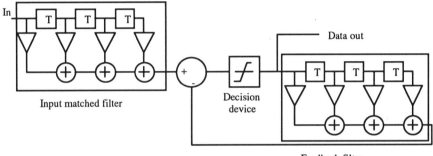

Figure 3.24   Decision feedback equaliser (DFE)

The complexity of a Viterbi decoder is proportional to the number of channel states, which is exponentially related to the length of the multipath spread expressed in symbol periods. Hence as the data rate increases, the complexity quickly becomes excessive. *Reduced-state sequence estimation (RSSE)* techniques have been developed to reduce the complexity of these techniques, at the cost of a small degradation in performance (of a few dB) [16]. These consider only the most probable states, or group states together. Of course, an RSSE with one state, reduces to a DFE.

Note that a wideband system contains a degree of inherent frequency diversity, since it is unlikely that the whole frequency range will fade at once. Hence, if a sufficiently effective equaliser is used (such as MLSE), a wideband system may perform better than a narrowband one. This is seen most clearly in CDMA systems, which usually occupy a much wider bandwidth than TDMA systems.

3.9   CDMA systems

3.9.1   The RAKE receiver

A CDMA system employs a code chip rate which is many times greater than the data rate, and may thus be classified as wideband. Usually, indeed, the signal bandwidth is several times the coherence bandwidth of the channel. This has two implications: firstly, the signal is subject to dispersion, and thus requires equalisation; and secondly, it exhibits the inherent frequency diversity mentioned above, and therefore can show advantages over narrowband systems. This form of diversity is sometimes known as *multipath diversity*.

The receiver structure used to implement the necessary equalisation and to realise the advantages of multipath diversity is known as the *RAKE receiver* (Figure 3.25) [31, 32]. 'RAKE' here is not an acronym; it likens the action of the receiver to a garden rake collecting the dispersed energy on its prongs and combining it together. In a conventional CDMA receiver, the received signal is correlated against a replica of the transmitted code sequence. This allows the signal to be recovered from noise and interference, which has no correlation with the replica code. In a RAKE receiver, the signal is fed to several correlators, each of which is fed with a delayed version of the code replica, corresponding to the channel delays, so that each synchronises with a different multipath component. The correlator outputs are weighted according to the component amplitudes, and corrected in phase, and then combined. In this way the multipath signals themselves are treated as diversity branches, of diversity with an order equal to the number of correlators.

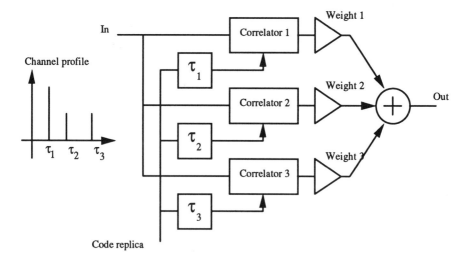

Figure 3.25   RAKE receiver for CDMA

3.10   Future developments

Development of mobile and personal communication systems continues apace, and this includes the modulation and coding aspects of the air interface. The primary goal of these developments, at least in cellular systems, is to increase user capacity for a given bandwidth, in terms of total user data rate per unit bandwidth per cell. There are two ways in which this may be done, both of which depend on modulation and coding. The first is to increase the spectral efficiency of the individual link, usually by the use of multilevel modulation techniques. The second is to increase the tolerance of a link to co-channel interference, so that reuse distance and cluster size may be reduced, increasing the overall system spectral efficiency. Coding is an important means by which this may be done. Coded modulation is a technique by which these two may be combined, bringing further advantages.

3.10.1 Multilevel modulation

The use of multilevel modulation [5] can increase bandwidth efficiency, because the number of bits transmitted per symbol is increased. As noted above, for an M point constellation there are $l = \log_2(M)$ bits per symbol. Since the bandwidth is determined by the symbol rate, the bandwidth efficiency is proportional to l .

However, there are two disadvantages to the use of multilevel modulation on a mobile radio channel. Firstly, as M increases, the spacing between constellation points decreases, and hence the link requires a greater signal-to-noise ratio. Mobile transmitters are of course subject to strict power limitations.   More seriously, it also becomes more vulnerable to interference, including co-channel interference. This may increase the required reuse distance, and so the overall result is no increase in system spectral efficiency.

Secondly, multilevel modulation is not usually constant envelope. While M-PSK schemes such as 8PSK (see Figure 3.11(c)) might in principle retain constant amplitude, in practice they are used in linear schemes, which do not. In any case, M-PSK beyond 8PSK is used very little, because the point spacing is so small. QAM schemes are therefore generally used for $M>8$ but they are inherently non-constant envelope.   Thus they are subject to severe distortions, of the sort illustrated in Figure 3.10, when used with the non-linear HPAs usually used in mobile radio.

Recently, however, there has been a great deal of work on linearisation of non-linear HPAs, while retaining their power efficiency. Two techniques are used: pre-distortion, where the transmitted constellation is distorted prior to amplification in such a way as to cancel the distortion due to the amplifier; and feedback linearisation, where the distortion is fed back around the amplifier so as to cancel it. Techniques such as these may allow multilevel modulation to be used even with power-efficient class C amplifiers.

3.10.2 Application of coding

As mentioned above, in current systems coding is used only to protect the most vulnerable bits of coded speech, and not to increase the general tolerance to CCI. In principle coding could allow further reductions in reuse distance, and hence reduce cluster size, even, possibly, allowing 100% reuse in some circumstances. Clearly, this would allow a three-fold increase in system spectral efficiency over the cluster size 3 used in GSM, apart from the additional redundancy required for the coding.

Further, current systems have to use the rate of transmission required for the worst-case interference conditions.   Interference in a cellular system varies randomly with time, and only occasionally reaches the worst-case level.   The RACE II ATDMA project proposes to make use of this using adaptive coding [38].   When the interference level is high, powerful codes are used to maintain error-free communication.   When the interference decreases, the code rate, the number of bits per code symbol, may increase, improving the spectral efficiency.

3.10.3 Coded modulation

The use of coding combined with multilevel modulation could overcome the disadvantage of multilevel modulation that it is less tolerant to interference, and the disadvantage of coding that it tends to increase the required bandwidth.   The concept of coded modulation was developed by Gottfried Ungerboeck and a number of other researchers in the early 1980s [37].   They showed that the coding and modulation need to be treated as a single entity, and the coding scheme designed with the modulation in mind.

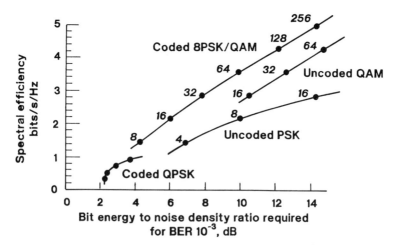

Figure 3.26   *Spectral efficiency versus required bit energy to noise density ratio for coded modulation compared with uncoded schemes*

Coding can then be incorporated into a system without increasing the required bandwidth by means of a higher-level modulation scheme. For example, uncoded QPSK (or π/4 QPSK) could be replaced by 8-PSK, together with a rate 2/3 code.

This would retain the same bandwidth efficiency while providing a significant coding gain: well in excess of 3 dB in this example. Equivalently, a multilevel modulation scheme could be used without suffering the penalty of reduced tolerance to interference, by exploiting the coding gain to allow the use of a more bandwidth efficient scheme overall. Thus, uncoded QPSK could be replaced by coded 16-QAM, increasing the bandwidth efficiency by 50%, while *increasing* interference tolerance by nearly 1 dB. Figure 3.26 compares uncoded and coded modulation schemes, showing the gains possible.

However, it has been shown that coded modulation schemes designed for the AWGN channel do not perform very well on the Rayleigh fading channel more likely in mobile and personal communication systems. Hence, considerable research effort has recently focused on the design of coded modulation schemes for fading channels, both Rayleigh and Rician. A well-designed scheme can give even greater coding gains than on the AWGN channel, because the coding introduces an implicit diversity effect. Thus coded modulation may readily be adapted for personal and mobile radio systems, and indeed is most likely to feature in any future TDMA systems.

References

1   Aulin, T, Lindell, G, and Sundberg, C-E., 'Selecting smoothing pulses for partial response digital FM' *IEE Proceedings*-F, vol 128, pp 237-244

2   Belfiore, C. A. and Park, J. H., 'Decision feedback *equalisation' Proceedings of the IEEE*, vol. 58, no. 5, pp 779-785, May 1970

3   *Bell System Technical Journal*, special issue on 'Advanced mobile 'phone service', vol 58, Jan. 1979

4   Bello, P. A., 'Characterization of randomly time-variant linear channels' *IEEE Transactions on Communication Systems*, vol CS-11, pp 36-393, Dec 1963

5   Borgne, M., 'Comparison of high-level modulation schemes for high capacity digital radio' *IEEE Transactions on Communications*, vol. COM-33, pp 442-449

6   Chuang, J. C-I., 'The effects of time delay spread on portable radio communications channels with digital modulation' *IEEE J. Selected Areas in Communications*, pp 879-889, June 1987

7   Clark, A. P., *Equalisers for digital modems* (Pentech Press, 1985)

8   Clark, A. P. and Brent, J. B., 'Narrow-band digital modems for land-mobile radio' *Journal of the IERE*, vol. 57, no. 6 (supplement), pp 293-303

9   Clark, G. C. and Cain, J. B., *'Error correction coding for digital communications'* (Plenum, New York, 1981)

10   Clarke, R. H., 'A statistical theory of mobile radio reception' *Bell System Technical Journal*, vol 47 pp 957-1000, July 1968

11   COST 207 *'Digital land-mobile radio communications'* Commission of the European Communities, 1989 (ISBN 0-7273-0504-2)

12   Cox, D. C and Leck, R. P., 'Correlation bandwidth and delay spread multipath propagation statistics for 910-MHz urban mobile radio channels' *IEEE Transactions on Communications*, vol COM-23, pp 1271-1280 November 1975

13   Couch, L. W., *'Digital and Analog Communication systems (3rd Ed.)'* (MacMillan, 1990)

14   Cox, D. C., 'Wireless Personal Communications: what is it?' *IEEE Personal Communications Magazine*, pp. 20-35, April 1995

15   Cox, D. C., 'Wireless network access for personal communications' *IEEE Communications Magazine*, pp 96-115, Dec. 1992

16   Eyuboglu, M. V. and Qureshi, S. U. H., 'Reduced-state sequence estimation with set-partitioning and decision feedback' *IEEE Transactions on Communications*, vol. COM-36, no. 1, pp 13-20, Jan. 1988

17   Farrell, P. G., 'Coding as a cure for communications calamities' *Electronics and Communications Engineering Journal*

18   Forney, G. D., 'Maximum likelihood sequence estimation of digital sequences in the presence of intersymbol interference' *IEEE Transactions on Information Theory*, vol IT-28, no. 3, pp 363-378, May 1972

19   Gilhousen, K.S., Jacobs, I.M., Padovani, R, Viterbi, A.J, Weaver, L.A, and Wheatley, C.E., 'On the capacity of a cellular CDMA system' *IEEE Transactions on Vehicular Technology*, vol. VT-40, no. 2, pp 303-312, May 1991

20 Glance, B. and Greenstein, L.J., 'Frequency-selective fading effects in digital mobile radio with diversity combining' *IEEE Trans. Communications*, vol. COM-31, pp 1085-1094, September 1983

21 Hamming, R. W., 'Error detecting and error correcting codes' *Bell System Technical Journal*, vol 29, pp 147-160, April 1950

22 Hodges, M.R.L., 'The GSM radio interface' *British Telecom Technology Journal*, vol. 8, no. 1, pp 31-43, Jan. 1980

23 ITU-R *'Provisional draft new recommendation on procedure for selection of radio transmission technologies for FPLMTS'* COST 231 TD(94)106, Darmstadt, September 6-8, 1994

24 Jakes, W.C. (Ed), *'Microwave Mobile Communications'* (Wiley Interscience, New York, 1974)

25 Lee, W.C.Y., *'Mobile Communications Engineering'* (McGraw-Hill, 1982)

26 Lin, S, and Costello, D. J., *'Error control coding: fundamentals and applications'* (Prentice-Hall, 1983)

27 Linnartz, J-P., *'Narrowband land-mobile radio networks'* (Artech House, 1993)

28 Murota, K. and Hirade, K., 'GMSK modulation for digital mobile radio telephony' *IEEE Trans. Comm*, vol COM-29, pp 1044-1050

29 MacWilliams, F. J. and Sloane, N. J. A., *'The theory of error-correcting codes'* (North-Holland, 1977)

30 Parsons, J. D. and Gardiner, J. G., *'Mobile Communication Systems'* (Blackie, 1989)

31 Proakis, J. G., *'Digital Communications (3rd Ed.)'* (McGraw-Hill, 1995)

32 Price, R. and Green, P. E., 'A communication technique for multipath channels' *Proceedings of the IRE*, vol. 46, pp 555-570, March 1958

33 Qureshi, S. U. H., 'Adaptive equalisation' *Proceedings of the IEEE*, vol 73, no. 9, pp 1349-1387, Nov. 1985

34 Rice, S. O., 'Statistical properties of a sine wave plus random noise' *Bell System Technical Journal*, vol. 27, no. 1, pp 109-157, Jan. 1948

35   Ross, A, and Zehavi, E., 'Propagation channel model for personal communications systems' *Proc. 44th IEEE Vehicular Technology Conference*, pp 185-189, June 1994

36   Sweeney, P., *'Error control coding: an introduction'* (Prentice-Hall, 1991)

37   Ungerboeck, G., 'Channel coding with multilevel/phase signals' *IEEE Transactions on Information Theory*, vol IT-28, pp 55-67, Jan. 1982

38   Urie, A., 'The ATDMA project' *Proceedings of Joint COST 231/227 workshop on 'Integration of satellite and terrestrial communications'*, Florence, April 1995 (Kluwer, 1995)

39   Viterbi, A. J., 'Error bounds for convolutional codes and an asymptotically optimum decoding algorithm' *IEEE Transactions on Information Theory*, vol. IT-13, pp 260-269, April 1969

40   Zhang, W., 'Simulation and modelling of multipath mobile channels' *Proc. 44th IEEE Vehicular Technology Conference*, p 160-164, June 1994

Chapter 4

Professional user
radio systems

Simon Cassia

Introduction

Professional user radio systems come in a very wide range of forms and complexity. They vary from simple 'infrastructureless' systems in which users with handportable or vehicle-mounted radios talk only to each other, to systems that are more complex and sophisticated than present day digital cellular telephone systems. In this chapter we will look at how and why these different systems have evolved, their structure and how they operate. We will see how digital techniques are coming into use in these systems, and will review briefly the Trans European Trunked Radio (TETRA) standard developed by ETSI for digital professional user radio systems.

In the following text the term 'mobile' radio has its conventional meaning of either a vehicle-mounted or a handportable radio. The widely accepted term 'private mobile radio' (PMR) is used to refer to these professional user radio systems, although several of them are in fact public systems on which users share access. More correctly, these latter systems are referred to as public access mobile radio (PAMR) systems.

4.1 What are professional user radio systems?

Radio systems that are currently in use throughout the world fall broadly into two categories:

- mobile telephone (cellular, CT2, DECT) systems, whose main aim is to provide a wireless extension to the public telephone networks; in these systems fundamentally the communication is between only two parties

- professional user radio (PMR) systems, whose main aim is to provide radio communications between a group of users engaged in some common activity; the group's activity is often managed by a central supervisor known as the dispatcher.

Typical users of PMR systems include public safety workers such as the police, the military, vehicle breakdown personnel, dispersed maintenance crews and building site workers.

There is significant overlap between the roles of mobile telephone and PMR systems. Groups who are engaged in a common activity can use cellular systems to communicate; e.g., some taxi companies dispatch their cabs using cellular phones. On the other hand, many PMR systems provide their radio users with connections to public telephone subscribers. However, there do remain fundamental differences between the types of service that professional radio users and mobile telephone users need, and the consequent characteristics of the systems they require. A selection of the important differences is listed in Table 4.1.

Table 4.1   Differences between the needs of professional users and mobile telephone users

Service/Characteristic	Professional User	Mobile Telephone User
Ability to operate in a group, with all members of the group hearing all conversations	Yes	Generally not required - limited capabilities can be provided by conferencing
Capabilities to allow dispatcher to control users	Yes	No
Call handling supplementary services (SS)	Tailored to each type of user and many PMR SS unlike telephone SS	Standard across entire system and similar to telephone SS
Press-to-talk type access to other users and the dispatcher, i.e. no need to dial	Essential	Generally not required
Near-zero call set up times	Essential for police, etc.	No
Scale of system	From single site to nationwide	Regional or national coverage
Very high capacity	Usually not required	Essential
Frequency planning	Often no frequency co-ordination with other nearby users	Frequencies co-ordinated across entire system
System provision	Often owned and operated by user, although there is significant use of shared (PAMR) systems	Services provided by system operator
Payment for the service	Pays for provision; use is 'free'	Pays for how much of the service is used

In summary, professional user radio systems are characterised by the greater range of services and features they provide, compared to the simpler, standardised set of services provided by cellular systems. As a result, these systems can be more technically complex than digital cellular systems and therefore can provide more of a challenge for the engineer.

4.2   PMR system operating modes

This section describes the two operating modes of PMR systems: conventional and trunked. Wide-area trunked operation is the most complex type of PMR system and so is given most attention in the rest of this chapter. However, in practice most radios in use in Europe operate in conventional mode, and of these most operate on single site systems.

4.2.1   Conventional operation

The conventional mode of operation of PMR systems is that the user turns on, or 'keys', the transmitter in the radio and then talks. Any user with a compatible receiver which is tuned to the frequency of the transmitter can then hear the transmitted speech. If the radio has no more sophistication than this (and many private mobile radios in use around the world do not) then the radio operates in 'open channel mode'.

Single channel conventional operation
In most developed countries PMR frequencies are strictly limited resources so regulatory authorities generally require organisations that generate small volumes of traffic to share the same frequency, or 'channel'. However, as each transmission is meant for only one of the group of users on the channel, it can be irritating for subscribers to hear every message transmitted. Additionally, for privacy or operational reasons, the dispatcher may want only one user to hear the message, so some method of identifying which subgroup or individual the message is for is required. Two systems are commonly used:

- Selective Calling (Selcall), in which each transmission is preceded by a burst of up to five audible tones that indicate the identity of the user the transmission is intended for; consequently, this system is also know as 'Five Tone'. There are three common standards for Selcall; CCIR , ZVEI and EEA. Tone frequencies lie between 700 Hz and 2800 Hz. Tone durations are 100 ms (CCIR), 70 ms (ZVEI) and 40 ms (EEA).

- Continuous Tone Controlled Signalling System (CTCSS), also known as Private Line (PL), in which a series of sub-audible tones between 67 Hz and 250 Hz are transmitted continuously with the speech. The exact frequency of the tones encodes the identity of the group that the call is intended for.

Radios in the group identified as the intended recipients of a message direct the audio to their speakers. Radios not in the recipient group keep their loudspeakers muted and lock their transmitters so that a user who is not in the recipient group and does not therefore hear the message cannot transmit over the top of the message. This feature is known as tonelock. To prevent one radio transmitting continuously and blocking all other users, each radio has a timer which prevents it from transmitting for more than a set period (normally a few tens of seconds).

One interesting use of Selcall is as a low speed data system. In many applications, the dispatcher wants to be kept updated about what each of its units is doing. Rather than using voice reports, the crew can simply indicate the unit's status (e.g. on the way to the incident, at the scene of the incident, on the way to hospital, back at base) by keying in one of a number of pre-defined codes to the mobile radio. The mobile then uses Selcall to transmit the code to the dispatcher's location . A computer system at the dispatcher site decodes the status messages and updates a display in front of the dispatcher. This is a very efficient use of spectrum, as a status message takes only about 500 ms to be transmitted, rather than the 10 s or so of speech required to pass the same information. Additionally, some radios provide automatic retries of status transmission, so the operator simply needs to enter the status code and the mobile will take care of waiting until the channel is free and transmitting the status code. Finally, it is much more efficient for the dispatcher, who does not have to enter manually the mobile's location or status into the tracking system.

Multiple channel conventional operation
If there are too many organisations using one channel it gets congested, despite the use of Selcall. To overcome this, the regulatory authority may allocate each organisation its own channel, or if its traffic is large enough, multiple channels. The simplest way for an organisation to use multiple channels is to allocate one to each different group of users in the organisation. Radios are fitted with a channel selector switch so that users can choose the group with which they wish to communicate. Users belonging to the same group can then communicate with each other without having to worry about interrupting other groups' communications. Many multichannel systems also use Selcall and CTCSS for providing privacy in individual calls, and for passing status messages.

4.2.2 Trunking operation

Conventional operation is not a very efficient way of using multiple channels, as one channel is allocated to each group even when none of the users in the group is talking. As shown in Figure 4.1, if there were 10 groups of users on a system (Groups A to J), and users in each group communicated for only 30% of the time, then for 70% of the time a channel dedicated to a group would be lying idle. A more efficient way of using the scarce radio frequencies is to allocate a channel to a group only when a user in that group wants to speak. Thus each time a user in the group wishes to speak, one of the free channels in the channel pool would be allocated to the group for just the duration of the conversation. At the end of the conversation the channel would be returned to the pool of free channels, ready to be allocated to a new group. This mode of system operation is known as 'trunking', and will be familiar to anyone with a cellular background, as it is the way that cellular systems operate.

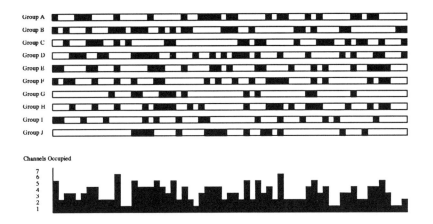

Figure 4.1   *Channel loading chart for 30% occupancy. With each group being busy only 30% of the time, six channels operating in trunking mode provide enough capacity to carry the traffic from 10 groups*

In practical conventional systems, the regulatory authorities would not allocate 10 frequencies as in the example above, but would allocate perhaps only six, and expect some of the groups to share the same frequency. The result of this arrangement would be that the groups sharing the same channels would block each other's calls. Trunking operation would allow these users to be allocated their own channel without blocking the calls of other groups and so would improve the system's utility.

In trunking systems one radio channel is normally allocated as a permanent control channel. The system uses this channel to receive call set up requests from subscriber radios and to inform other radios of the frequency of the radio channel that has been allocated to each call (we will see later that the control channel is used for many other purposes as well). Allocating a channel for control purposes introduces inefficiencies, because that channel can not be used to carry traffic[1].

---

[1]   In practice it is possible to allocate the control channel for use as a traffic channel during times of congestion; the penalty is that the system can no longer signal to radios not already involved in a call. The result is that features such as priority call set up and some data services

Despite this factor, trunking systems are usually better for users who wish to have many groups in their system, because trunking allows many groups to be carried by few channels, whereas conventional operation would require a channel for each group.

Trunking systems allow groups of users to operate independently in the same way as conventional systems, but rather than being associated with a particular frequency each group is identified by a group number. Users normally select the group to which they wish to communicate by using a control on the radio. Some systems provide a scanning facility, so that radios can monitor the conversations of other groups when their own group is quiet.

4.3 Optimising single site coverage

Unless special technical measures are taken, the coverage provided by each site in the system will not be optimal because of the following two issues.

i) *Path imbalance:* there is a fundamental imbalance in radio systems, in that the power of the base station transmitter is normally several dBs higher than the power of the mobile radios. This imbalance is particularly noticeable for handportable radios, which typically operate between 1 - 4 W whereas base stations typically operate between 10 - 25 W in Europe. Outside Europe this imbalance can be even greater as base stations often operate at significantly higher transmit powers. The result is that the system has a much greater range outbound, i.e. base to portable, than inbound, i.e. portable to base.

ii) *Poor mobile coverage:* PMR system users generally require that all members in a group can hear the transmissions of any other member of the group. There are a number of factors which make direct communications between mobiles more difficult than between the base and the mobiles:

- the base station antenna is located high up in a position which gives good coverage over the area; in contrast, two mobiles might be in dips in the ground from which they can each see the base antenna but not each other's antennas

- if a mobile is positioned on the opposite side of the base station from the transmitting mobile (as is the case for mobiles A and B in Figure 4.2) then it will be beyond the range of the other mobile

- to reduce some of the transmission power imbalances mentioned earlier, the base receive antenna may be designed with additional gain; the mobile

---

cannot be provided while the control channel is temporarily allocated for carrying traffic.

stations do not have this benefit, so while a base station may be able to receive a mobile successfully, another mobile located the same distance away as the base station may not.

The following sections describe potential technical solutions to these issues.

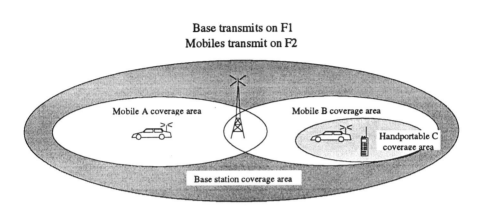

Figure 4.2   *Mobile ranges are normally less than base station ranges. The base station can act as a repeater, rebroadcasting messages received on frequency F2 on frequency F1 so that all users can receive them. Vehicle mounted radios can also act as repeaters to provide coverage for low power handportables*

4.3.1 Balancing inbound and outbound range

Two techniques are commonly used in PMR systems to balance the ranges:

Increased base receive antenna gain
Base receive antennas can be designed to have higher gain than the transmit antenna so as to improve inbound performance, either through using a different antenna type or by using multiple antennas, a technique known as diversity reception. The effectiveness of diversity reception depends very much on the separation between the antennas measured in wavelengths, and so tends to be effective only at frequencies above UHF. However, provided that the power imbalance is not too great, these techniques are capable of providing a completely balanced link.

Satellite receivers
As shown in Figure 4.3, satellite receivers can be located towards the edge of the coverage area to receive the signals from mobiles which would be too weak to be received by the base. In this arrangement a voting system is needed at the central site to select the best of the received signals. Several different approaches are used:

- the audio from each receiver is summed and the total signal used; this approach is non-optimal because the signal strengths at some of the receivers will be low, resulting in a poor overall signal to noise ratio for the summed signal

- the system can examine each receive path and select the one with the highest power

- the system can examine each receive path and select the one with the best signal to noise ratio, for example by selecting the signal with the lowest power at the upper edge of the signal pass band, on the assumption that this power is almost entirely noise.

This third option has some benefits over the second as it is more resilient to noise or interference which could appear as a strong signal in the signal pass band, and it also allows for the effects of signal to noise degradation in the landlines connecting the satellite receivers to the central voting unit.

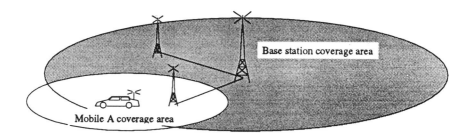

Figure 4.3   *Satellite receivers near the edge of coverage are able to receive the weaker signals from the mobile and thus provide symmetrical inbound and outbound coverage*

4.3.2 Improving mobile-to-mobile communication

To alleviate the mobile-to-mobile communication problem, systems often use repeaters that receive any signal transmitted on the receive frequency and re-transmit them, usually at a higher power, on the transmit frequency. This makes it possible for users who do not have direct contact with a transmitting mobile to receive its message. There are four common types of repeater.

Base repeaters
Base stations themselves often act as repeaters, re-transmitting any signals they receive, as well as relaying the received signal to a dispatch operator located in a control room. For this reason, in the PMR world base stations are often referred to as repeaters.

Community repeaters
In many countries where there is widespread use of open channel mode radios with no dispatcher, 'community' repeaters are often installed. These devices provide the range extension for mobile radios communication within the area served by the community repeater. Community repeaters often use Selcall or CTCSS to prevent unauthorised radios from accessing the service the repeaters provide. Each repeater is set to respond only to a limited range of codes, so that it repeats only the signals from those radios which use the correct code.

Transportable repeaters
Some organisations, particularly public safety, use small portable repeaters which can be carried around to improve communications at, for example, a traffic accident.

In-vehicle repeaters
Many organisations use vehicles to carry staff to the scene of an incident and then the occupants leave the vehicle and use handportable radios for communications. These vehicles are often fitted with repeaters which receive the signals from the vehicle's crew when they are using their handportable radios and relay the signals to the base station. A common approach is to operate handportables on UHF frequencies with the link back to the base station being at VHF.

4.4 Providing wide area coverage

There are both legal and practical reasons which limit the amount of power that can be radiated by transmitters, and so a single site can cover only a limited area. To provide wide-area systems, multiple base station sites must be used; however, the same methods mentioned above for optimising single site coverage can also be used in multi-site systems.

As shown in Figure 4.4 there are broad methods by which frequencies are assigned to the sites:

• different frequencies at each site, termed cellular

• the same frequency at each site, termed quasi-synchronous, often shortened to quasi-sync, and also called simulcast.

4.4.1 Cell-based frequency assignment

In this scheme, each site is allocated a different group of frequencies to create a cell of coverage. The mobiles must change frequency as they move from cell to cell. There are three main methods by which this frequency change is controlled.

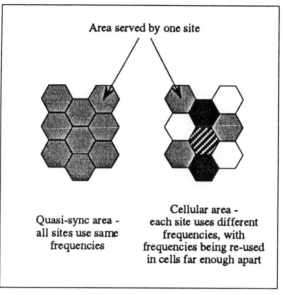

Area served by one site

Quasi-sync area -
all sites use same
frequencies

Cellular area -
each site uses different
frequencies, with
frequencies being re-used
in cells far enough apart

Figure 4.4   *Frequency allocations in quasi-sync and cellular frequency assignments*

i) *Manual selection*: in this scheme each user takes responsibility for deciding which base station serves the area in which it is operating, and selects the appropriate frequency manually. This method works well when the area served by each base station is relatively large and users do not change location too often.

ii) *Scanning*: in this scheme the frequency allocations to the base stations are the same as in the manual selection scheme, but the mobile radio is loaded with a list of the frequencies assigned to the various sites in the system. The radio scans the frequencies in this list until a signal with acceptable strength (or bit error rate in digital systems) is found. If the signal strength falls below a pre-defined limit, the mobile re-scans until another suitable signal is found.

iii) *Scanning with location update*: in the first two schemes the system is not aware of the location of each mobile, and so must either transmit the same message at all sites or the subscriber must accept that the contents of the transmission will be different in different cells. An example of this situation is when each region is served by a different dispatcher, and the user communicates with different dispatchers as it moves through the coverage area. In more sophisticated 'scanning with location update' systems, the infrastructure and mobile negotiate together to select the best frequency at any time and the infrastructure then directs the communications for each subscriber to only the cell in which it is then located. This scheme can very considerably improve the efficiency with which frequencies in the system are used, as there is now no need to transmit messages on frequencies in cells where there are no subscribers wishing to hear them. The price to be paid is a much more complex system, with a sophisticated over air signalling protocol to enable the cell selection and location updating communications and a very intelligent central control unit to keep track of the location of each user and to direct the correct communications to each user's cell. Scanning with location update is the wide-area coverage scheme used in cellular telephone systems.

Manual selection and scanning systems normally operate in conventional mode. Because of the potential for using frequencies in cells only as needed, scanning with location update systems normally operate only in trunking mode.

The frequencies used at the base stations in a cell can be used at base stations in other cells provided there is sufficient separation between the cells so that they do not interfere with each other. The basis of the reuse plan is the cell repeat pattern; this is the collection of cells and frequency assignments which can be repeated across the coverage area. Figure 4.4 shows a 4-cell repeat pattern, although 7-, 9- or 21-cell repeats are more commonly used. Cell repeat patterns are covered in more detail in other chapters.

4.4.2 Quasi-synchronous operation

In this scheme, each site in the quasi-sync area is allocated the same frequencies, and on each frequency the same message is transmitted from every site. As the mobile radio moves around within the area served by the quasi-sync sites it does not need to change the frequency it is tuned to nor does it need to do any handoff as it passes from the coverage provided by one site to that of the next. In effect, the additional sites make it appear as if the coverage of a single site has been extended over a much wider area. Either conventional or trunking mode can be used within the quasi-sync area.

The coverage areas of the sites are designed so that there is a substantial overlap. This overlap achieves two things:

- it improves dramatically the quality of the coverage at the edges of the site coverage areas; for example, if the system was designed for 90% probability of coverage at the site coverage area boundaries, then a mobile which was on the edge of the coverage area of two cells would have something like a $(1 - (1 - 0.9)^2) = 0.99$ probability of coverage

- it provides some 'fill-in' coverage if a site fails.

The main elements of a typical quasi-sync system are shown in Figure 4.5.

Systems with quasi-sync frequency assignments are much more complex to engineer than systems with cellular frequency assignments. Special measures must be taken to:

- equalise the outbound signal launch times

- handle the severe RF interference that the mobiles experience from the multiple signals they receive

- design the areas in which strong signals overlap to minimise delays between the signal arrival times

- select the optimum inbound signal.

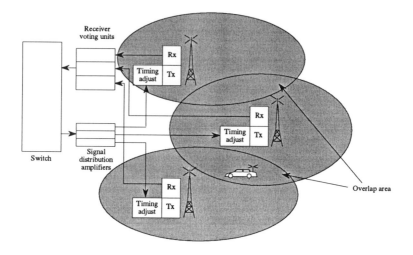

Figure 4.5   *Quasi-sync systems need timing adjustment at the transmitter sites to equalise the signal launch times and voting units to pick the best combination of signals from those received from the mobiles*

Equalising the outbound signal launch times

To achieve quasi-sync operation the signals from each transmitter covering the overlap area must arrive at the mobile at almost the same time or there will be destructive interference between the baseband elements (e.g. the audio) of the signals.   Destructive interference of the radio frequency components is unavoidable, and is discussed later.   Experience indicates that the baseband components of the signal must be received with a phase difference of no more than 30° of each other or the quality of the baseband signal will be unacceptably degraded.   Taking a typical maximum voice band frequency of 3 kHz, this requirement means that signals from different transmitters must be received with no more than 20 μs delay between them.   To meet this rather demanding objective, the baseband signals must be launched from every base station at the same time, and the propagation times must differ by no more than 20 µs. As these signals will be carried over circuits from the central site which will each impose a different delay on the signal's timing, compensation equipment must be installed at each base station site.

RF carrier interference
Destructive interference in the overlap area between the RF carriers from different sites is inevitable. To the mobile, it appears as very bad multipath interference. Typically, the receiver may be faced with two signals of equal power, but with a delay of up to 20 µs between them, and with other signals from other transmitters up to 40 dB below the primary signals and with delays up to 100 µs. This is a demanding environment for a receiver to operate in, and demands an equaliser in digital systems which have high symbol rates.

If the transmit frequencies of the quasi-sync transmitters were exactly synchronised, in principle a stable RF interference pattern would be set up across the coverage area, with areas where the destructive interference from two equal strength transmitter signals resulted in a permanent null. In practice, this effect is most pronounced at VHF (particularly low band); at UHF the wavelengths are much shorter and so natural disturbances in the environment tend to make the nulls in the interference pattern move around. The presence of fixed nulls is acceptable when the nulls are in areas where coverage is not required, for example, on top of hills. However, normally the effect is removed by offsetting the transmitter frequencies from each other so that the interference pattern changes with time and no location in the coverage area suffers a permanent null. The value of the offset depends on the frequency and modulation scheme in use and ranges from a few hertz to a few tens of hertz.

Overlap area
The importance of the overlap area is shown in Figure 4.6. In areas A and C, the signal strength of the distant transmitter is far enough below that of the main signal so that the mobile receiver will lock on to only the main signal, by the FM capture effect[2]. In these areas the mobile is not receiving quasi-sync signals.

---

2   Because of the need to track the frequency variations which form the basis of FM, these receivers tend to lock to the stronger of two signals, provided there is sufficient difference between the two signal strengths. This phenomenon is called the capture effect and typically occurs when there is between 8 and 12 dB static difference in the two largest signal strengths when these signals are towards the limit of sensitivity of the receiver, although this ratio depends very much on the receiver's design. For strong signals the difference between the two signal strengths is much less. In dynamic conditions, i.e. when the signal is varying due to fading, an extra margin of about 8 dB needs to be added.

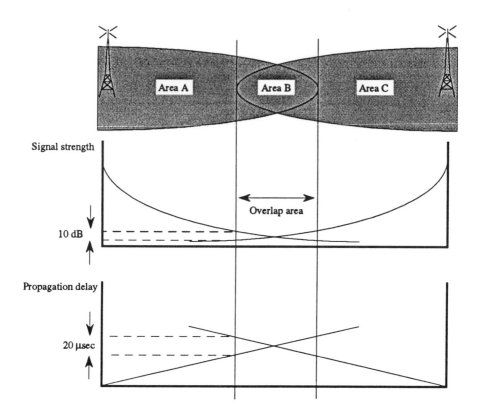

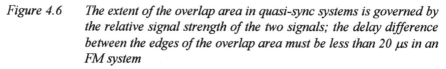

Figure 4.6   *The extent of the overlap area in quasi-sync systems is governed by the relative signal strength of the two signals; the delay difference between the edges of the overlap area must be less than 20 μs in an FM system*

In area B the strengths of the primary and secondary signals are close enough for the FM capture effect not to work. Here there is interference between the two baseband signals after the RF signal has been demodulated, so the system must be designed to ensure that the maximum difference in delay between the two sides of overlap area is less than the 20 μs specified above. This limits the width of the overlap area to 6 km, namely 20 μs × the speed of light.

In practice the overlap area is likely to be designed to be less than this value, for the following reasons:

- the signal strength difference at which the capture effect begins to operate, and thus the relative strengths of the signals that define the overlap area, cannot be known very precisely due to differences in receiver design

- within the overlap areas the delays of the signals may be quite variable due to multipath reflections.

On top of these issues, there are likely to be other regions of the coverage area where the signals from two base stations may have sufficiently similar levels to prevent the receiver capturing only one of them. These regions can often be close to one of the transmitters; for example, if the terrain shadows one of the transmitters, or if the terrain tends to favour the propagation of signals from a remote transmitter. Thus, in practical systems, very careful design of the radio coverage is required to reduce to a minimum the extent of extraneous overlap areas and to keep the dimensions of the planned overlap areas well within the delay limits that the mobile receivers will tolerate.

In digital systems the situation is more complex, as the capture effect now depends on the digital modulation scheme and the acceptable delay differences for the chosen symbol[3] rate. Taking typical values for the TETRA system (discussed later in this chapter) in which 10 dB static is required for capture, the symbol rate is 18 ksymbol/sec and a symbol overlap of 0.25 symbol is tolerable[4] before significant symbol errors start to occur, the theoretical width of the overlap area being about 4 km. However, the practical constraints mentioned above are likely to reduce this significantly.

Selecting the inbound signal
On the inbound path, several base station receivers capture the signal from a mobile, and the system must choose the combination of these RF signals that provides the best baseband signal. The same voting schemes as are discussed in Section 4.3.1 are used.

4.4.3 Comparison of scanning with location update and quasi-sync operation

Because quasi-sync systems appear to the mobile radios as one large site, no special features or functions are required in them for handoff between sites and so quasi-sync is an attractive way of extending coverage without needing the more complex 'scanning with location update' cell based radios. Additionally, the system does not need to keep track of the location of each mobile and so the

---

3   In digital systems the rate at which the modulation pattern of the RF signal changes is called the symbol rate. In many modulation schemes each symbol can encode more than one bit; for example, in TETRA, the transmitter has a choice of sending one of four different symbols during each symbol period, and so each symbol encodes two bits.

4   The tolerable symbol overlap can be extended by using an equaliser in the receiver.

infrastructure control is less complex than an equivalent cell-based arrangement, although it does need timing units and voting equipment. Finally, because of the overlap between adjacent sites, in principle it is easier and cheaper to provide high quality coverage over the whole coverage area with quasi-sync than with an equivalent cell-based system. However, designing a quasi-sync system is much harder than a cell-based system, and once in operation, the quasi-sync system may need much more 'fine tuning' before it works satisfactorily.

In cell-based systems, as a subscriber passes from one cell to another during a call, it must re-tune its radio to the correct frequency in the new cell. Because of this hard handoff (the mobile is either communicating with one cell or the other, but never both) the coverage improvement in the overlap areas experienced by quasi-sync systems is not realised in cell-based systems.

Some systems take advantage of both types of operation, by operating in a cell-based mode, but with each cell being a quasi-sync area served by several sites. Properly designed, systems like this can make most effective advantage of the benefits offered by the different wide area coverage schemes.

4.5 Wide area trunking system operation

Wide-area trunking systems are the most complex in the PMR world. They can employ all the techniques so far discussed above in the single site and wide area coverage sections to provide high quality mobile-to-mobile and mobile-to-base communications across a wide area. This section provides a basic description of how these sophisticated systems operate.

4.5.1 Typical wide area trunking system architecture

In trunking systems users are allocated radio channels only when they wish to send a message to the other members of the group. To enable this functionality the system requires a sophisticated control infrastructure as well as a switching infrastructure which can configure the one-to-many type connections that are needed if all the members in a group are to hear the message being sent by the transmitting user. The architecture of a typical trunking system is shown in Figure 4.7.

At least one radio channel is dedicated for use by control signals. In a frequency division multiple access (FDMA) system it would be a frequency; in a time division multiple access (TDMA) system it would be one of the time slots on one of the TDMA frequencies. The signalling on the radio channel is always digital; signalling rates vary but typically lie between 1200 b/s and 9600 b/s, the higher bit rates being required both to increase the capacity and hence the sophistication of the signalling and also to reduce the call set-up times.

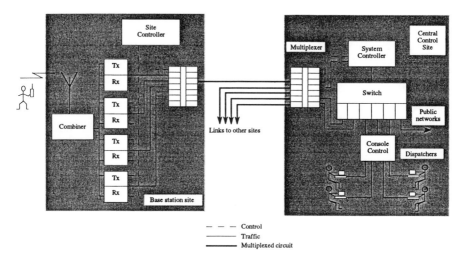

Figure 4.7   Architecture of a typical trunking system

Within the network, control and traffic channels are normally logically separated but are sometimes combined on the same intersite links, using multiplexing. This measure is made more easy when traffic within the network is carried digitally, even though it may be either analogue or digital on the radio path. Systems that use analogue speech on the radio path may keep the speech analogue throughout the network or may convert it to digital either at the base station or at the switch. If so, the industry-standard 64 kb/s Pulse Code Modulation (PCM) is often used, because components and software which work to this standard are readily available.

Figure 4.7 shows that switching and control are centralised. In practice, these elements may either be concentrated in one place or may be distributed around the network. In the same way, dispatch consoles and access to the public networks may either be carried out from a central site or distributed around the system. The dispatcher console normally has a special status in a PMR system. It may be able to alter the members which form each group, define which members of a group may transmit and which must operate in receive only mode, and limit the access of individual radios to the public networks.

4.5.2 Setting up a group call

The basic operation of a multi-site trunking system is to set up a call from one member of a group to the other members, who will normally be located in several of the cells in the system. The system will also set up individual calls (i.e. from one mobile to another), and interconnect calls (from a mobile to the public networks).

Group calls are invariably half duplex in nature, but individual calls may be either half duplex or full duplex, which is equivalent to a call on a cellular telephone system. Some systems will allow the call to be used for data as well as for voice. The operation of all these call types is a sub-set of the group call set-up described below.

Mobiles generally initiate a call by pressing the PTT switch on their radio (individual calls may be set up using a key pad to dial the outgoing number and waiting for the called radio to respond). This action causes the radio to send a message over the control channel to the system control, requesting that a call be set up from it to all other members of the group. The control channel is a common resource shared by all the radios in the cell, and so the calling radio must follow the access method defined by the system for sending messages on the control channel. This procedure may introduce a delay if, for example, the channel is already being used by another radio, or another radio contends for the control channel at the same time as the calling radio. The slotted ALOHA protocol is often used as the random access method.

Once the system controller receives a call set-up request from a member of a group, it must carry out a number of activities:

- find out which cells contain the other members of the group

- find a free channel in each one of these cells, and in the call originator's cell

- tell all the users in the group the frequencies of the channels which have been allocated for the call in their particular cells

- configure the switching infrastructure to connect the channel used by the call originator to the sites where members of the group are located; note that in general this will be a one-to-many connection

- signal to the call originator that it may proceed.

In a typical system users are looking for all these activities to be completed in about 500 ms, so the performance required of the system controller is very stringent indeed.

Most systems allocate both an inbound and outbound radio path for the group call in each cell. Allocating both directions of the radio path for group calls is inefficient, as this type of call always operates in half duplex mode, i.e. a mobile is only ever transmitting or receiving in a group call, but never both simultaneously. Independent trunking, in which the system allocates inbound and outbound channels to different calls, makes more efficient use of the radio channels but requires much more sophisticated channel allocation and control, and is not in common use.

4.5.3 Radio channel trunking

A group call will be initiated by a member of the group, and then other members of the group will respond to the message. In all, there are likely to be a number of transmissions from different members of the group before the call is completed. The system can either allocate different radio channels for each transmission, or it can allocate and hold the channels for the entire call.

The first approach is called transmission trunking and makes the most efficient use of the radio channels, but places a very high load on the system controller because of the extensive signalling it must carry out with the MSs each time a transmission begins and ends. The second approach is called message trunking and relieves the controller of any further configuration of the radio paths once they have been established for the call, although it has to reconfigure the connections within the switch each time a new member of the group wants to talk. Neither of these approaches is ideal and most systems use a half-way house called quasi-transmission trunking. In this approach, the radio channels are not deallocated immediately the transmission ceases (as in transmission trunking) but are held for a period known as the hang time. If another user in the group begins transmitting within the hang time the channels are not deallocated, and another hang time will begin when that user stops transmitting.

The common control channel is always used for the initial call request message and the initial channel allocation messages to the mobiles. After the mobiles have retuned to the traffic channel, all further control information must be sent in the traffic channel. A number of strategies exist for sending signalling in the traffic channel without disturbing the traffic in the channel. These include both in-band and out-of-band signalling (using similar techniques to CTCSS) and, in digital systems, time slot stealing in which one or more time slots are stolen from the digitally encoded speech and used for signalling. The digital voice decoder is notified of the slots which have been stolen and smoothes over discontinuities in the speech path.

4.5.4 Other control channel activities

The control channel is busy with many activities other than call control:

- as users move from cell to cell they must update the controller about their whereabouts so that it knows to where to send incoming calls; this process is known as registration

- as mobiles move across the boundaries of a cell they must agree with the infrastructure which cell they will move into

- while not making calls, mobiles may be instructed to enter a battery saving mode by the infrastructure

- before allowing a mobile to register on the system, the central controller may carry out an authentication process to make sure that the mobile is authorised to use the system

- in some systems the control channel is used to carry data and status message traffic between users.

4.6 PMR radio frequency environment

Having considered the technical aspects of PMR systems, we must turn our attention to another critical aspect of these systems; the licensing and regulatory conditions under which they operate.

4.6.1 PMR spectrum

PMR systems operate in many bands depending on the country and the type of user. For example, most countries reserve specific bands for police/fire/ambulance communications. The approximate frequencies of the main PMR bands in Europe are VHF low band (68 - 88 MHz), VHF high band (146 - 174 MHz), VHF Band III (UK only: 174 - 225 MHz), UHF (420 - 450 MHz and 450 - 470 MHz) and TETRA (380 - 400 MHz). Frequency spacings are normally 12.5 kHz, 20 kHz or 25 kHz, depending on the system deployed.

Generally speaking, users prefer VHF frequencies for systems operating in rural areas because they believe that the system range is greater at these frequencies than at the higher frequencies due to the increased propagation loss at UHF compared to VHF. However, there are a number of factors which in practice equalise the range capabilities of the two frequency bands:

- there is less man-made noise at the higher frequencies, so signal to noise ratios at UHF degrade less slowly with distance (measured in wavelengths) than at VHF

- the most efficient antennas have dimensions which are of the same order as the signal wavelength; because of the shorter wavelength at UHF, antennas of a practical size can be made to be more efficient than those at VHF

- in hilly areas the inherent better propagation of VHF provides no advantages over UHF because both systems are limited by line-of-sight considerations.

4.6.2 Frequency co-ordination and adjacent channel power

With a few notable exceptions, e.g. between Germany and Benelux, there is little co-ordination of frequencies between users or even between administrations in different countries in Europe. The most striking example of this is that, although

UK and continental police forces use the same frequencies, the allocation of frequencies to base station and mobiles is reversed. The result of this arrangement is that high power transmissions from continental base stations sometimes swamp the lower power transmissions from the UK mobiles, causing unacceptable interference. There are various co-ordinating bodies for spectrum across Europe, but the allocation of spectrum has a long and complicated history so it will be a long time before there can be harmonisation of frequencies. The one exception, however, is TETRA, where most European countries have allocated the spectrum from 380 - 400 MHz for public safety TETRA users.

Because it is impractical to co-ordinate spectrum use between PMR users, the Adjacent Channel Coupled Power Ratio (ACCPR)[5] for PMR systems is very tight. The specification against which PMR radios are licensed, European Technical Specification (ETS) 300 086, requires at least 70 dB isolation between the on-channel and adjacent channels in FM systems with 25 kHz channel spacing, and 60 dB isolation at 12.5 kHz channel spacing. This tight specification is needed to allow two radios which are operating on adjacent channels to work satisfactorily when they are close to each other and distant from the base station. Cellular systems can operate with much lower ACCPRs because the system designers make sure that adjacent channels are not used in adjacent cells. Consequently, a typical ACCPR in cellular systems is 30 dB.

4.7 Digital PMR systems

In common with most communications systems, PMR systems are gradually making the move to digital. In fact, for many years several systems have used digital voice schemes within the fixed part of the network, but on the air interface the majority of systems are still analogue FM. However a number of manufacturers either already supply systems or are planning systems which use digital speech on the air interface.

4.7.1 Benefits of digital PMR systems

Most current PMR systems operate using FM with channel spacings of either 25 kHz or 12.5 kHz. While there have been discussions about lower channel spacings of either 6.25 kHz or even 5 kHz, and quite novel schemes have been

---

5   The ACCPR is a measure of how much the transmitter 'splatters' power into channels other than the one in which it is transmitting. A receiver in one of the adjacent channels of a 1 W mobile operating with an ACCPR of 60 dBc would receive only 1 mW of interference power. The receiving radio could thus receive signals from a 30 W base station which was 50 times further away than the interfering transmitter.

proposed for achieving these channel spacings, it is generally accepted in the industry that digital modulation schemes provide a better way of reducing the effective channel bandwidth.   Modulation schemes are in use which allow significant bit rates in 25 kHz bandwidth; TETRA provides 36 kb/s and iDEN (integrated Digital Enhanced Networking), another technology currently proceeding through the ITU standards process, uses a modulation scheme which gives bit rates of 64 kb/s in 25 kHz. Given that bit rates of 7 kb/s (including error correction coding) are perfectly feasible for digital speech, digital systems allow four and even six speech channels to be carried on a single 25 kHz channel.

Digital systems allow the receiver to detect and correct errors in the received bit stream by using the forward error correction coding embedded within the stream. With the addition of substantial error protection (about an additional 50% b/s on top of the raw speech for TETRA) the performance of the system at the edges of coverage can be very substantially improved. Thus the quality of speech is constant over the whole coverage area, rather than degrading as the distance from the base station increases, as is the case with analogue FM systems. The error correction capability of the digital signal also makes the speech signals much more resilient to the multiple radio paths often encountered in PMR systems.

As the radio path is designed to handle bits, it makes carrying data over the radio path much more simple than with an analogue radio path. Integration of information services (voice, data, video, image etc.) is therefore a practical proposition on digital systems. Encryption, which is becoming a requirement for most public safety services, is also much easier to provide in a digital system than in analogue ones.

4.7.2 Digital modulation schemes

To implement a digital radio system a modulation scheme must be chosen which transmits bits over the radio path. This subject has received very substantial study in recent years, and so many schemes have now been proposed that one is almost spoilt for choice.

An important objective in selecting the modulation scheme is to maximise the channel capacity, which means maximising the ratio of bits per second to hertz at a given signal to noise ratio[6]. Classes of modulation scheme in which both the signal's phase and amplitude are varied provide the highest values of bits per second per hertz[7], and hence channel capacity. This combined phase and

---

[6]   It is always easy to increase the available b/s/Hz by improving the signal to noise ratio, but this reduces the range of the signal from each site, as signal to noise ratio is a function of distance from the transmitter.

[7]   Some digital modulation schemes, for example Gaussian minimum shift keying as used in GSM and 4-level FSK, do not have any amplitude modulation on the carrier. However, these schemes provide lower b/s/Hz at equivalent signal-to-noise ratios than schemes such as QAM and

amplitude modulation is in marked contrast to FM, where only the signal's phase varies. Filtering the output of a digital modulator to keep the transmitted signal within the allocated channel bandwidth can also result in amplitude modulation of an otherwise phase-only modulated signal, such as QPSK.

Some quite simple trigonometry shows that any non-linearities in the transmitter result in this amplitude modulation mixing with the carrier and causing additional small transmitted signals at various frequencies, as shown in Figure 4.8. Some of these signals lie in the adjacent channels, and so cause interference with any radios operating in these channels. Because these signals lie close to the wanted signal, it is not possible to filter them out with any practicable filter. The magnitude of this phenomenon, known as splatter, depends directly on the non-linearity in the final Power Amplifier (PA). As Figure 4.8 shows, splatter is not a problem in so-called constant envelope schemes in which the amplitude of the modulation does not change, because the signals resulting from the non-linearities in the PA fall well away from the wanted signal and so can be filtered out relatively easily.

Unfortunately, all PAs contain some non-linearities, i.e. as shown in Figure 4.9, a graph of input signal against output signal is not quite a straight line. The high efficiency class C PAs used in FM radios display very marked non-linearities; these devices provide no output at all until the input signal crosses a certain threshold and are extremely non-linear after that.

To allow amplitude-varying signals to be carried in narrow bandwidths without causing unacceptable splatter, the radio needs a PA which is very linear. However, linear PAs bring some drawbacks:

- amplifiers with high linearity are complex to implement. The basis is usually a moderately linear PA circuit, with additional linearising circuitry added; these additional components require additional space and consume extra current from the battery.

- linear amplifiers are much less efficient than the class C amplifiers currently used in FM radios; i.e. for a given RF output power the class C amplifier draws much less power from the battery than would a linear amplifier. The main reason for this is that simply to achieve the linearity requirements, the PA must operate low down its characteristic curve (nearer the horizontal axis in Figure 4.9), where the device is more linear. At these low output power levels the device's efficiency is low.

- because the PA device is operating low down its characteristic curve, a device with a much higher rated output power must be used than if the device were

---

and QPSK, which do have amplitude modulation (in QPSK amplitude modulation is introduced by the band pass filters which limit the bandwidth of the modulated signal to the channel bandwidth).

operated in class C. As a result, the amplifier may be larger and more expensive.

•   linear amplifiers are less efficient than equivalent class C amplifiers, and so produce more heat which must be dissipated in the radio's chassis.

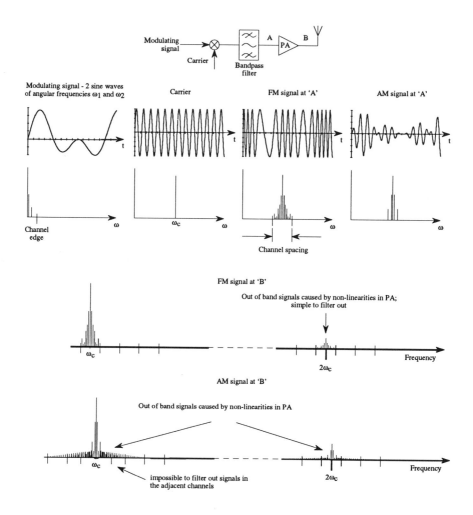

Figure 4.8   *The amplitude modulation of the carrier causes splatter in the adjacent channel which cannot be filtered out. The splatter in constant envelope FM signals is so far away from the wanted signal that it can be filtered out easily at the output of the PA*

As always, there are trade-offs to be made when choosing the modulation scheme. Most companies have chosen to implement non-linear schemes (i.e. schemes without AM) in their first generation radios but second generation products generally use linear modulation schemes. One of the prime movers here is the advent of TETRA, which uses π/4 DQPSK together with a tight specification for ACCPR (60 dB in the adjacent channel), and thus requires a very linear transmitter.

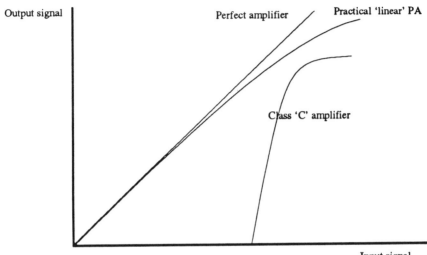

Figure 4.9   *Non-linearities in amplifiers; all amplifiers lose linearity as the input signal increases; Class C amplifiers are designed for high efficiency and as a result are highly non-linear*

4.7.3 Speech in digital PMR systems

To minimise the bit rate required on the radio path, digital PMR systems generally use vocoding techniques rather than analogue-digital conversion of the speech waveform, such as is used in PCM and ADPCM. Vocoding techniques make use of the relatively slow varying nature of the sounds which make up speech. Although typical band-limited speech may be formed from frequency components up to 3.4 kHz, the rate at which the loudness, pitch and tonal balance of the speech vary is only a few tens of hertz at the most. In fact, the rate at which the characteristics of the speech vary is limited by how quickly the speaker changes the shape of the vocal tract (the throat, tongue and lips) which shape the sounds. Vocoder operation is quite complex in concept, and is beyond the scope of this chapter. However, an analogy is that rather than transmitting the variations in the speech waveform itself, vocoders work by transmitting information about the basic pitch of the speech and the shape of the vocal tract, and this allows the receiving end to reconstruct the speech. As these characteristics vary quite slowly, it is feasible nowadays to transmit acceptable quality speech at bit rates below 5 kb/s. Work is underway to allow toll quality speech (i.e. as good as on the public telephone network) to be encoded at 8 kb/s.

Vocoded speech has two main limitations:

- an entire sample of speech must be collected before it can be processed; typically a few tens of milliseconds of speech is encoded at once. As the encoding and decoding process itself takes a significant amount of time, quite significant delays are built into vocoded systems even before normal transmission delays are included.

- generally, a speech signal which has been vocoded once cannot be vocoded again, as the reduction in quality becomes intolerable.

4.7.4 TETRA PMR standard

In 1989 the European Telecommunications Standards Institute (ETSI) began work on developing a digital PMR trunking standard for use by European organisations. Named TETRA (Trans European Trunked Radio), it is the European-side standard for digital PMR, analogous to the GSM standard for digital cellular telephony.

Interface standards
The TETRA standard defines eight interfaces:

- the trunked mode air interface between mobiles stations (MS) and the SWitching and Management Infrastructure (SwMI), which comprises the base stations, switches and controllers

- the direct mode air interface, for direct communication between MS without involving the SwMI

- the inter-system interface between two TETRA SwMIs

- an interface for connecting a network management terminal

- the interface to a line connected terminal, which has the same features and functions as a MS, but is connected to the SwMI by a landline, rather than by radio

- the terminal equipment interface between a MS and an external data device, such as a printer or video camera

- the interfaces between the SwMI and other networks, such as the public telephone network, public data networks and PABX networks.

One important point to note is that, unlike GSM, there are no interfaces defined with the SwMI[8]. ETSI took this decision as it did not want to constrain the range of architectures that manufacturers might want to choose to implement the system. The trunked mode air interface itself has two variants: the voice and data (V+D) interface and a Packet Data Optimised (PDO) interface. The main difference is in the framing of the bits on the radio path. In the V+ D version, the 25 kHz radio channel is subdivided into four subchannels using TDMA, and each subchannel would be allocated to a circuit mode connection with a basic capacity of 7200 b/s. In the PDO version the entire bit stream is available to the radio for the duration of its transmission, and there is no circuit mode type operation. As the V+D standard is the system of interest to the majority of users, in this chapter we will now concentrate only on this interface. However, in most respects the PDO air interface uses similar concepts for cell reselection and message transfer.

TETRA V+D services
The TETRA V+D standard provides the following services over the air interface:
- circuit mode connections such as half duplex group call, simplex broadcast call, full duplex individual call and full duplex calls to external networks such as the public telephone network, providing:
- voice at 4.6 kb/s using the algebraic code excited linear predictive (ACELP) codec algorithm

- data at 7.2, 14.4, 21.6 and 28.8[9] kb/s without forward error correction (FEC)

- data at 4.8, 9.6, 14.4 and 19.2 kb/s with medium levels of FEC

- data at 2.4, 4.8, 7.2 and 9.6 kb/s with high levels of FEC

- 30 supplementary services which provide additional functionality on top of the basic circuit mode services noted above. Some examples are:

- call authorised by dispatcher, in which the dispatcher is able to verify and approve a call request before it is allowed to proceed

- priority call, in which the infrastructure gives priority access to calls which have been sent with priority status

- discreet listening in which an authorised user may listen to one or more communications between TETRA subscribers without any indication to any subscriber that the communication is being monitored

- packet mode data connections:

---

8   GSM defines two internal interfaces: the A and A bis interfaces.

9   The data rate provided depends on how many of the four TDMA timeslots are allocated to the call.

- a connection-oriented network service (CONS) which is similar to the OSI network service provided by the X.25 protocol

- a specific connectionless network service (S-CLNS), which is similar to the service provided by the Internet Protocol (IP)

• a short data service (SDS), which is capable of passing free text messages containing up to 256 characters and up to 65 536 status messages.

TETRA V+D air interface

The V+D interface provides the lower three layers of the OSI 7-layer reference model. Two of the layers are further subdivided. The function of the layers is described in Table 4.2, and the overall structure of the interface is shown in Figure 4.10.

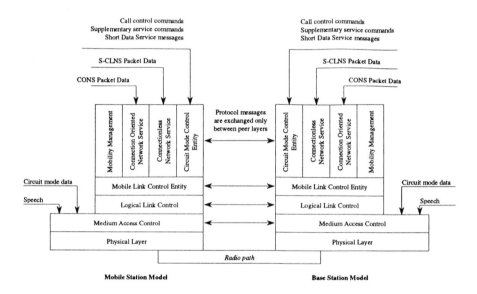

Figure 4.10   TETRA mobile and base station functional models

Characteristics of the physical and MAC layers

To complete this section we will take a slightly more detailed look at the format of the radio aspects of the TETRA air interface, which is defined fully in ETS 300 392 Part II.

TETRA operates with a channel spacing of 25 kHz. The modulation scheme used is $\pi/4$ DQPSK, as used by US Digital Cellular. A root raised cosine filter with roll-off factor of 0.35 is used to filter the signal to a bandwidth of nominally 18 kHz. With these parameters, the symbol rate is 18 ksymbols/sec, and as each symbol encodes 2 bits the gross bit rate is 36 kb/s.

Table 4.2   *Functions of the TETRA V+D air interface layers*

OSI Layer	TETRA Sub Layers/ Functional Entities	Function
Network	Mobility Management (MM)	In the SwMI: keeps track of which users are currently active on the system and their location. In the MS: controls which cells the MS is allowed to roam into and decides whether the MS is successfully connected to the network. MM also allows the MS to switch in to modes which consume less power when the MS is not involved in a call
	Circuit Mode Connection Entity (CMCE)	Controls the signalling for establishing circuit mode calls between the MS and the SwMI. Also controls the radio channel trunking during the call (as explained earlier in this chapter). Provides the short data service
	Connection Oriented Network Service (CONS)	Provides the control and signalling for the connection-oriented packet mode data service
	Specific Connectionless Network Service (CLNS)	Provides the control and signalling for the specific connectionless packet mode data service
	Mobile Link Control Entity (MLE)	Shields the network layer entities above it from the disruptions in communications caused when the MS changes cell. The MLE in the MS measures the signal strengths in adjacent cells, and using this and other information broadcast by each cell decides which cell it will move into
Data Link	Logical Link Control (LLC)	Provides two classes of data transfer capability to the MLE: • basic link in which each frame is treated as a separate communication, although frames may be acknowledged and there is error control • advanced link with error control, retransmission, flow control and sequence numbering so that frames are guaranteed to be delivered free from error and in the correct order
	Medium Access Control (MAC)	Provides the 4:1 TDMA format on the radio path and packs the messages received from the LLC into the appropriate slots. Also carries out the forward error correction, scrambling and interleaving (techniques used to reduce errors). Controls access to the random access control channels Note that this is the highest level in the stack that circuit mode traffic encounters; above the MAC, circuit mode data traffic is routed directly to the receiving application; voice traffic is routed to the ACELP codec and from there to the audio circuits
Physical	Physical	Uses the bit stream provided by the MAC to modulate the carrier for the transmitted signal; demodulates the received signal and passes the resulting bit stream to the MAC

As mentioned previously, the ACCPR is 60 dBc, which requires a very linear transmitter, in the order of 55 dB IMR3[10].

Three classes of receiver are specified: operation in normal urban and rural terrain, operation at high speed in hilly terrain and operation in quasi-sync. The dynamic sensitivity of the receiver in the normal urban terrain is -103 dBm for 2.2% bit error rate for both the mobile and handportable.

The 36 kb/s bit stream is subdivided into frames of 4 time slots or channels, each time slot containing 510 bits. The time slot duration is 14.167 ms. The net bit rate of each channel is about 7.2 kb/s once bits for receiver training, phase adjustment, PA linearisation (uplink) and broadcast information (downlink) have been accounted for.

One channel per base site is normally allocated for random access signalling from the mobiles; a slotted ALOHA access protocol is used. The time slots in the 18th frame of all traffic channels are allocated for channel associated signalling, and other time slots within the traffic channels may be 'stolen' to carry further channel-associated signalling.

Acknowledgements

I would like to thank David Chater-Lea of Motorola's LMPS Technology Centre and Gary Aitkenhead, Clive Jervis and Steve Valentine of Motorola's European Research Laboratory for their help in preparing this chapter.

---

[10]   Third order intermodulation products from a standard two-tone test are 55 dB below either of the tones.

Chapter 5

Cellular radio planning methods

Stanley Chia

Introduction

Radio planning is a key step for cellular network design and implementation. It is a multi-dimensional optimisation exercise based on many constraints, and includes spectrum allocation, traffic forecast and infrastructure investment. The planning is expected to meet a number of criteria, including a high user capacity, large coverage area, a high quality of service and a minimum quantity of network elements. Radio planning is dependent on many factors, such as the multiple access technique adopted by the system, regulatory and environmental constraints and user behaviour. In general, the overall objective is to provide a large coverage footprint and high user capacity.

The exercise requires extensive and detailed input information on the spectrum allocation for the service area (national spectrum allocation) as well as neighbouring areas (international co-ordination). In addition, it requires terrain, morphology and demographic data of the corresponding areas. While radio planning tools with sophisticated radiowave propagation prediction models are useful to assist radio planning, the accuracy is dependent on the topology of a country and may not always yield reliable results. The primary output of radio planning is to provide search circles for base site acquisition as well as frequency plans for channel assignment. Extensive field tests may be required to confirm the feasibility of the base sites. In order to validate the network plan, network optimisation has to be performed after the implementation of a base site.

This chapter covers the key aspects of radio planning and provides insights into the practicalities of the planning process. Examples of GSM planning are considered throughout in order to convey the important concepts as well as the trade-offs.

5.1 Spectrum utilisation

The allocation of radio spectrum can have a fundamental impact on the success of any cellular system. More spectrum would generally imply less expensive network infrastructure investment and a better operational efficiency. Although it is true that by reducing the cell size, a denser frequency reuse pattern may be achieved, this is, however, highly system dependent. The first step in radio system design is to confirm the amount of spectrum allocation to be awarded to an operator. Then with other system parameters, such as speech coding technique and requirements for data transmission, the amount of radio network equipment can be estimated.

For GSM the spectrum allocation at 900 MHz is categorised into the primary GSM band and the extended GSM band. Both bands support full duplex transmission using two sub-bands spaced 45 MHz apart. The standard, or primary GSM band, runs from 890 MHz to 915 MHz on the up-link (mobile transmits, base receives), and 935 MHz to 960 MHz (base transmits, mobile receives) on the down-link. The extended GSM band runs from 880 MHz to 915 MHz on the up-link and 925 MHz to 960 MHz on the down-link. This is summarised in Table 5.1.

Table 5.1 GSM spectrum allocation

	Extended GSM Up-link	*Extended GSM* Down-link	*Primary GSM* Up-link	*Primary GSM* Down-link
Start	880 MHz	925 MHz	890 MHz	935 MHz
Stop	915 MHz	960 MHz	915 MHz	960 MHz

In countries where analogue TACS systems exist, the primary GSM band may be subdivided further into a TACS sub-band and a GSM sub-band. The former occupies the lower 15 MHz of the 25 MHz duplex allocation, while the latter occupies the upper 10 MHz duplex band. For the primary GSM band, the radio frequency channels (carriers) are numbered from 1 to 124 and the corresponding frequency can be found from the following equations:

Mobile transmits $\quad\quad F_{up}(n) = 890.2 + 0.2\,(n\text{-}1)$ MHz

Base transmits $\quad\quad\quad F_{down}(n) = F_{up}(n) + 45$ MHz

For the DCS1800 band, the frequency band runs from 1710 MHz to 1785 MHz on the uplink, and 1805 MHz to 1880 MHz on the downlink. Full duplex transmission is supported by using two sub-bands spaced 95 MHz apart. Similar to the primary GSM band, the radio frequency channels for the DCS1800 band are numbered from 512 to 885 and the corresponding frequency can be found from the following equations:

Mobile transmits $\quad\quad F_{up}(n) = 1710.2 + 0.2\,(n\text{-}512)$ MHz

Base transmits $\quad\quad\quad F_{down}(n) = F_{up}(n) + 95$ MHz

With a guard band of 200 kHz at each end of the subbands and the radio frequency channel spacing of 200 kHz, a maximum of 174 carriers is allowed in the extended GSM band and 374 carriers in the DCS1800 band.

For a start-up GSM network, a minimum of 12 carriers is normally required in order to meet the carrier-to-interference ratio (CIR) for satisfactory quality of service. However, most regulators allocate 24 or more carriers to an operator, so that two carriers can be implemented within each cell.

If more than one operator is assigned to the same GSM band, one guard channel is normally the minimum separation requirement to avoid interference between two adjacent operators. International coordination of the frequency allocation is also required at national borders. This should normally conform to CEPT TR 20/08 and CEPT TR 25-04, which provides guidelines for the harmonisation of frequency usage and planning at international boundaries. Specifically, CEPT TR 20/08 states that the frequency coordination in border areas is based on concepts which may be summarised as follows: preferential frequencies or preferential frequency bands may be used without coordination with a neigbouring country, if the field strength of each carrier produced by the base site does not exceed 19 dBμV/m, at a height of 3 m above ground, at a distance of 15 km inside the neighbouring country. All other frequencies are subject to coordination between administrations if the interfering field strength produced by the base sites exceeds 19 dBμV/m, at a height of 3 m above ground, at the border between two countries.

Apart from these matters, there are three key areas where greater spectrum efficiency can be achieved in a cellular network: better trunking efficiency, tighter frequency reuse distance and a higher density of base sites. It is, however, important to note that the efficiency of a network is ultimately dependent on the amount of spectrum available.

5.1.1 Better trunking efficiency

A higher user capacity can be achieved through improving the trunking efficiency of the system. This can be achieved in two ways: more spectrum allocation and low rate speech coding.

Additional spectrum

More available spectrum implies that more transceivers can be added to individual base transceiver stations that will improve the trunking efficiency. Note that one base site may consist of one or more base transceiver station (BTS); one BTS corresponds to the equipment for one cell. The gain of capacity through better trunking efficiency is in excess of the increase in spectrum i.e. the process is not simply linear, see Section 5.2 below.

Half-rate coding
Although half-rate coding doubles the number of voice circuits for a transceiver, the overall network capacity gain may not always be fully achieved. The reason for this is that the utilisation of half-rate coding within a network is also determined by the number of full-rate users (such as roamers and subscribers with full rate mobile equipment) and the amount of 9.6 kbit/s and 4.8 kbit/s data transmission as well as facsimile transmission, within the network. All these users will require full rate channels.

5.1.2 Reduce inter-site distance

Reducing the inter-site distance, or cell radius, and increasing the number of base sites, can lead to tighter reuse of the frequency sets and hence a higher user capacity. There are, however, penalties to pay and limitations as explained below.

Increase in infrastructure
Clearly an increase in infrastructure will increase the cost of the network and the need for placement of more antenna structures within the area. This will not only render the network more expensive, requiring more maintenance effort, but will also incur more challenges on environmental issues and difficulties in base site acquisition.

Quality of service
The quality of service for a GSM system is defined by a set of measurements which includes call success rate, time to connect services, handover success rate and bit error ratio for data services. While GSM proposes a handover success probability of 99%, this only holds true under 'low traffic conditions' [GSM ETR 02.08 Section 3]. Degradation is expected to increase for congested base sites. With a reduction of cell radius and a constant average mean call holding time, it is likely that the number of handovers will increase. In particular, in high traffic demand areas, where the cell radius has to be decreased, congestion is also likely to occur. In this case the cumulative handover success probability will be severely compromised leading to a general degradation in the quality of service for the network.

Handover delay
Following each handover, mobile stations have to re-construct their neighbour cell list before a further handover can be initiated. This requires a finite duration and could be a problem if the cell radius is reduced to below a certain extent. In this situation, mobile stations will not be able to initiate another handover in time so that they can be connected to the best server base site, and hence the overall carrier-to-interference ratio of the network will be degraded.

Switch loading

With small cell radius, more frequent handover could lead to added loading on the base station controller. A base station controller is the dedicated network equipment for managing the radio resources. Its functional responsibilities include handling of the mobile station connection, radio network management, base transceiver station management and traffic concentration. Evidently, the dimensioning of the base station controller has to take into account the volume of handover traffic. A larger number of base sites, therefore, translates to additional network element cost and reduced system efficiency.

5.1.3 Tighter frequency reuse

The frequency reuse pattern is dependent on the carrier-to-interference ratio that a system can tolerate. For a GSM system, a 4 x 3 reuse pattern is normally required for satisfactory operation in an ideal environment. Insufficient spectrum could lead to problems with frequency planning especially when the implementation of irregular cell size is needed in a real environment.

Slow frequency hopping

In principle, the introduction of slow frequency hopping leads to additional gain in the link-budget under both noise and interference limited situations. However, the gain is environment and mobile speed dependent. The highest gain is seen when a large number of carriers is available and in urban environments with the mobile station either stationary, or moving very slowly. Even then, the gain is only a few dB. For this reason, it is unacceptable for a radio planner to generally assume that with slow frequency hopping, a 3 x 3 reuse pattern could be adopted, especially when the spectrum allocation is limited. For a good quality network, it is necessary to design the network based on a nominal 4 x 3 reuse pattern.

Interference management

It is not always feasible to implement a regular 4 x 3 frequency plan in a real world using a minimum of 12 frequency groups. This is especially difficult in areas where a base site is missing in the reuse pattern, due to difficulties in site acquisition, or in areas where there is a change in cell size. Extra frequency groups have to be implemented in order to maintain the carrier-to-interference ratio. For coastal regions and bay areas, where the base sites are exposed to each other and over water paths exist, extra frequency groups are frequently required to ensure the overall network quality. Without adequate spectrum, a high quality of service is difficult to engineer in areas where a regular reuse pattern is disturbed.

Discontinuous transmission

Discontinuous transmission is a further feature of GSM that can be employed to reduce the interference level in the network. It employs a voice activity detector to differentiate between speech and silence periods. As there is no information transmitted during the silence period, the transmitter can be switched off over that duration. In an interference limited system, discontinuous transmission will introduce a carrier-to-interference ratio gain by reducing the overall level of interference. Assuming the voice activity cycle is about 50%, a theoretical carrier-to-interference ratio gain of up to 3 dB can be achieved by using discontinuous transmission in a mature GSM network. The exact value is dependent on the level of background noise and the specific implementation.

5.2 Network dimensioning

The number of channels available for voice and data communications determines the capacity of a radio network. An understanding of the traffic theory is, therefore, essential for the design of a cellular system. The amount of traffic generated by each subscriber is a key input in deciding the number of channels required. There are two common traffic formulae used for cellular radio traffic engineering: Erlang B and Erlang C. The former represents the situation where blocked calls are lost, while the latter represents the situation where blocked calls queue for an indefinite duration until a channel is obtained. When calculating the required number of channels based on the amount of offered traffic, the blocking probability, or grade of service, has to be defined. When the Erlang B formula is used, a 2 % blocking probability is commonly adopted, while a 1 % blocking probability is normally employed in association with the Erlang C formula. It should be noted that the Erlang C formula is more conservative in terms of traffic capacity. An example is shown in Table 5.2.

Table 5.2 Traffic capacity estimation

Number of channels	Traffic capacity	
	Erlang B, 2% blocking	Erlang C, 1% blocking
6	2.28	1.76
14	8.20	6.71
22	14.0	12.5

In the design of a radio network, the traffic forecast in a geographical location has to be distributed amongst the base sites to be deployed. Depending on the terrain of the area, the best server region for each base site could be different. In this situation, the amount of traffic captured by each cell will not be the same. It is, therefore, important that the coverage prediction and network dimensioning are well integrated.

In the real world, traffic distribution is rarely uniform. A uniform traffic distribution can only be approximated for large cells in rural areas. In city centres, the traffic normally peaks and gradually reduces towards the outskirts. There may still be local peaks in suburban centres and motorway junctions. For cell dimensioning, it is a usual practice to optimise the network by using the same number of channels, but changing the cell coverage area so that the traffic carried per cell is kept constant with the traffic density. This optimisation exercise can normally be achieved by controlling the base station transmit power, antenna footprint and up-link/down-link balancing parameters.

The introduction of half-rate speech coding will increase the overall capacity of the network. In theory, the gain in capacity will be in excess of twice that of full-rate speech coding due to the trunking efficiency. However, as explained in Section 5.1.1, in a real network, there are factors that affect the trunking efficiency of half-rate speech coding. These factors include the amount of full-rate equipment in the network due to roamers and early subscribers and the amount of data service that requires full rate transmissions, e.g. facsimile. Furthermore, the ability of the base station system to pack dynamically and optimally the half-rate channels will also have an impact on the overall efficiency of half-rate speech coding. Clearly, network dimensioning has to take all these factors into account.

The dimensioning of the control channels is more complex, because there are many different services on the control channels, some having a variance in their duration while the others are deterministic. As short message services utilise the stand-alone dedicated control channel, a non-combined control channel structure, i.e. the stand-alone dedicated control channel assigned on a different time slot with respect to the common control channel, is preferred in urban areas, in anticipation of the potential volume of short message services traffic. A combined control channel structure is, however, preferred in remote areas in order to maximise the traffic capacity of a base site.

5.3 Link budget design

The radio link between the mobile station and the base transceiver station is best described by a link budget. This budget shows the system gains and losses and is determined by many factors which include:

> base station transmit power
> base station receive power (base station reference sensitivity)
> mobile transmit power
> mobile receive power (mobile reference sensitivity)
> base station antenna gain
> mobile antenna gain
> base station diversity reception gain
> path loss
> base station combiner loss
> base station feeder / connector loss
> frequency hopping gain (see Section 5.1.3 above)
> building / vehicle penetration loss (for in-building/in-vehicle coverage)
> operating margins.

An example of a simplified link budget is presented in Table 5.3.

Table 5.3 Example of a radio link budget design for GSM Class 4 mobile stations

Receiving end	Units	Base transceiver station	Handportable
Reference sensitivity	dBm	-104	-102
Interference degradation margin	dB	3	3
Cable/connector loss	dB	4	0
Receive antenna gain	dBi	12	0
Lognormal margin	dB	5	5
Isotropic power needed	dBm	-104	-94
Transmitting end			
Transmit power	dBm	33	38
Combiner loss	dB	0	3
Cable loss	dB	0	4
Transmit antenna gain	dBi	0	12
Peak effective isotropic radiated power	dBm	33	43
Isotropic path loss	dB	137	137

Because cellular radio operates with a duplex link between the base transceiver station and the mobile station, the maximum distance for communication is determined by the weakest link. For handportables (Class 4 mobile stations), the output power is 2 W. A base transceiver station can, however, operate at a power level an order of magnitude higher. For this reason, it is necessary to balance the up-link transmission budget with that for the down-link. This is particularly important for two-way communication near the cell edge. Both speech and data transmissions are dimensioned for equal quality in both the up-link and down-link directions. As the same type of antenna is used for both the transmit and receive path and they are installed close to each other, both the antenna gain, cable and connector losses and the path loss are assumed to be the same for the up-link and the down-link directions.

What are different are the transmit power levels and the receive reference sensitivity. In addition, on the down-link, there is always a disadvantage of combiner loss if more than one transceiver is used. In the up-link direction, the output power of the mobile station is limited to 2 W for handportables, for safety and battery life considerations. There are, however, advantages in terms of diversity gain in the up-link due to the implementation of base station spatial antenna diversity. For a balanced up- and down-link, one can write

Base station transmit power – combiner loss – mobile receiver sensitivity
= Mobile transmit power + diversity gain – base station receiver sensitivity

As the physical configuration at different base sites may be different and the mobile transmit power is constant, it can be concluded that the effective radiated power will be different for each base site. This is somewhat in contradiction with the equal effective radiated power for each base site in order to enable path loss handover implementation. It is therefore important to realise that perfect link budget balancing can only be achieved in an ideal world. In a real world, where path balancing is not always achievable, a down-link limited network is preferred to an up-link limited network.

Finally it should be noted that link budget balancing is only achieved for a specific power class of mobile station. When more than one power class of mobile station exist in the same network, link budget imbalance is taken care of in cell selection in idle mode, and in the handover decision algorithms, in most situations.

5.4 Radio coverage planning

Radio coverage is frequently perceived to be the most important measure for network quality. Clearly the extent of the coverage footprint is directly proportional to the size of the network infrastructure investment. As for quality of service requirements, the coverage requirement for any network is a marketing decision. For GSM application, one of the fundamental criteria is that the network provides adequate service to handportables.

From an operator point of view, the primary goal of providing radio coverage for a cellular network is to achieve a high traffic capacity while maintaining an acceptable quality of service. The provision of radio coverage to remote areas, where the traffic is minimal, is frequently seen to be a poor investment and is provided purely for strategic reasons.

The design criterion used for coverage of a cell is to meet a 90 % location probability within the service area. The signal levels received at both the mobile station and the base transceiver station have to meet the threshold specified in GSM Technical Specification 05.05. These levels are referred to as the reference sensitivity. For the base transceiver station this is -104 dBm, while for the handportables it is -102 dBm. In order to ensure reliable communication, the planning figures used for radio planning have to include an additional margin to account for the shadow fading. The margin is dependent on the standard deviation of the received signal level and the path loss characteristic. In addition, dependent on the coverage objective, a suitable building penetration loss margin will also be required. This margin is highly dependent on the type of building material and the orientation of the building relative to the base site. For on-street coverage, a margin is again required for vehicle penetration. This is again dependent on the type of vehicle as well as the orientation of the vehicle relative to the base site. A compensator or booster (commonly known as car kit) can be installed to compensate for the loss. The power class of the mobile station has to be maintained, however.

For GSM systems, cells can generally be classified into large cells and small cells. For large cells, the base station antenna is installed at prominent positions on top of tall buildings, and the path loss is determined mainly by diffraction and scattering at roof tops in the vicinity of the mobile station. The nominal cell radius could be well in excess of 3 km especially for rural area implementation. For urban areas, small cell coverage is normally adopted. The antenna elevation in this situation is normally above the median, but below the maximum height, of the surrounding roof tops. The cell radius implemented may be less than 1 km. In addition, microcells with the base station antenna mounted below the city sky line can also be implemented. Micro-base station classes are defined in GSM Technical Specification 05.05. The cell radius is in the region of 200 - 300 m. The implementation of a micro-base station has to take into account the minimum coupling loss between the mobile station and the base transceiver station. This is determined by the relative antenna positioning, gain and height.

For designing radio coverage to special areas such as tunnels, underground railway systems, city underpasses, etc., it is necessary to consider the dimensions of the environment. In particular, it is important to decide if dedicated equipment, including repeaters or dedicated base stations, have to be implemented. In some extreme cases such as train tunnels, it is necessary to use leaky coaxial cable to provide adequate radio coverage. On-frequency repeaters are frequently used in GSM systems in order to provide adequate signal for strategic coverage. These repeaters are essentially bi-directional amplifiers that extend the footprint of the service area by amplifying the up-link and the down-link signals simultaneously. High power repeaters are used at rural areas, while low power repeaters are mainly for indoor applications. In real life implementation, there are many planning considerations, such as coordination among operators, as well as radio frequency engineering issues such as the derating of the repeater gain to meet the GSM intermodulation specification.

5.5 Cellular architecture

The principle of a cellular network is based upon the reuse of the same frequency set. The architecture of the cells is determined by the user capacity of the network to be delivered by a given frequency allocation as well as the consistency with the overall quality of service. At launch, the primary objective of a cellular network is to maximise the radio coverage footprint, thereby extending service to as many geographical areas and customers as possible. As the network evolves, more base sites are added to increase the capacity and to enhance coverage to selected areas.

Reusing an identical radio carrier in different cells is limited by co-channel interference between base sites. In order to ensure that the carrier-to-interference ratio is met, the nominal reuse distance employed must be sufficiently large. By contrast, the smaller the reuse distance, the higher the traffic capacity. It is, therefore, a trade-off between these two requirements. For a practical network, there will be a mixture of single-BTS (omni-directional) base sites, or multiple-BTS (sectorised) base sites, to be deployed. A single-BTS base site is fed from a single group of transceivers into one set of transceiving omni-directional antennas, while a multiple-BTS base site has more than one group of transceivers at the same base site each with its own transceiving antennas.

For GSM implementation, there are a number of possible cell repeat patterns that could be adopted. If 12 frequency groups are available, a reuse distance of 12 would ensure a 12 dB location reliability over at least 90 % of the service area. For multi-cell BTS base site implementation, a reuse distance of four with three cells (4 x 3), or a reuse distance of three with three cells (3 x 3), may be selected. In general, a 4 x 3 reuse pattern is recommended due to a better overall carrier-to-interference ratio location reliability.

Reducing the reuse distance from 4 x 3 to 3 x 3 will increase the user capacity. It is potentially possible to implement a 3 x 3 configuration for GSM. However, in order to achieve an acceptable quality of service, it is necessary to ensure that sufficient carriers are available at each base site for the implementation of slow frequency hopping. In addition, power control and discontinuous transmission will also have to be implemented and optimised. It can be seen that the frequency set reuse distance is dependent upon the number of frequency groups N. The larger the frequency group, the greater is the co-channel reuse distance D. If R is the radius of the cells, simple hexagonal geometry of the cells will give the following relationship:

$$D/R = \sqrt{(3N)}$$

Two typical reuse patterns for GSM are shown in Figures 5.1 and 5.2.

For conventional GSM systems, a single layer of cells with different cell radii is normally adopted. At network start-up, base sites have to be deployed at relatively prominent locations so that a reasonably large coverage footprint can be achieved with a minimum amount of infrastructure investment. As the network traffic grows, small cells, or microcells, can be deployed to increase the overall network traffic capacity.

Alternatively, a more flexible approach to meet the network evolution requirement is to adopt a mixed cell architecture as traffic hot spots develop. With a mixed cell architecture, macrocells are used for wide area and background coverage, while the underlaying microcells are used for providing strategic coverage to traffic hot spot locations. The microcells can also be deployed to provide coverage to localised poor reception areas.

5.6   Frequency planning

RF channel allocation is normally made on a frequency division basis. A group of neighbouring cells (a cluster) uses the same set of carriers, but there is no reuse within the same cluster. Due to the carrier-to-interference ratio requirement, the number of frequency groups for a cluster in a GSM implementation is either 9 or 12. Channel allocation for uniform traffic distribution may follow one of the well-known reuse patterns depending on carrier to interference ratio requirements. By

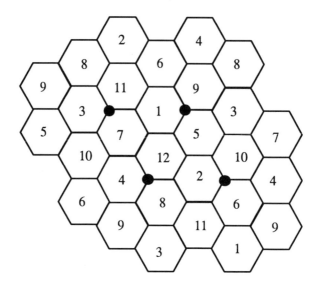

Figure 5.1   Frequency plan for a 4 x 3 reuse pattern

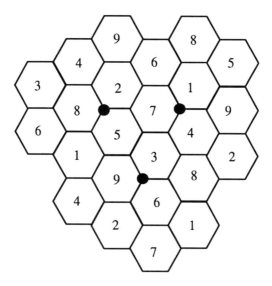

Figure 5.2   Frequency plan for a 3 x 3 reuse pattern

contrast, channel allocation for non-uniform traffic distribution can be optimised using graph colouring heuristics. Furthermore, for irregular terrain and highly built-up areas, non-standard frequency assignment may have to be made. It should, however, be noted that irregular frequency planning could lead to severe problems when a new base site is inserted into the existing network. The affected areas and hence the number of base sites involved could be quite extensive.

Adjacent channel interference is an additional consideration in the frequency assignment. According to GSM recommendations, the adjacent channel suppression should be -9 dB. This implies that the first adjacent channel should not be used in the same cell or the same site. It should be noted that when a 3 x 3 reuse pattern is used, it may not be possible to avoid adjacent channel interference, as it is impossible to avoid adjacent frequencies in neighbouring cells.

As an example the frequency allocations for a 3 x 3 and a 4 x 3 reuse plan are shown in Table 5.4. With this assignment we can see that the carriers are always 9 and 12 carriers apart at the same cell, for the 3 x 3 and 4 x 3 reuse plans, respectively, thus avoiding adjacent channel interference by carriers within the same cell. The same principle applies to other regular reuse patterns

Table 5.4 Frequency allocation schemes

Cell number	1	2	3	4	5	6	7	8	9
Frequency group	A1	B1	C1	A2	B2	C2	A3	B3	C3
Carrier number	f_1	f_2	f_3	f_4	f_5	f_6	f_7	f_8	f_9
	f_{10}	f_{11}	f_{12}	f_{13}	f_{14}	f_{15}	f_{16}	f_{17}	f_{18}

a) RF channel allocation for a 3 x 3 reuse pattern

Cell number	1	2	3	4	5	6	7	8	9	10	11	12
Frequency group	A1	B1	C1	D1	A2	B2	C2	D2	A3	B3	C3	D3
Carrier number	f_1	f_2	f_3	f_4	f_5	f_6	f_7	f_8	f_9	f_{10}	f_{11}	f_{12}
	f_{13}	f_{14}	f_{15}	f_{16}	f_{17}	f_{18}	f_{19}	f_{20}	f_{21}	f_{22}	f_{23}	f_{24}

b) RF channel allocation for a 4 x 3 reuse pattern

In addition, with two carriers-per-cell spectrum allocation, it is possible to implement 'upper/lower band' frequency planning at the early phase of network roll-out. The principle is to alternate, where possible, the allocation of the broadcast control channel (BCCH) carriers and the traffic channel (TCH) carriers in adjacent clusters. This could maximise the co-channel reuse distance of the broadcast control channel carriers that are constantly transmitting at full power. The effectiveness of this method will gradually diminish as the network attains maturity. The frequency allocation scheme for the 'upper/lower band' implementation is shown in Table 5.5.

Table 5.5   *An 'upper/low' band frequency allocation plan for a 4 x 3 reuse pattern*

Cell number	1	2	3	4	5	6	7	8	9	10	11	12
Upper band												
Frequency group	A1	B1	C1	D1	A2	B2	C2	D2	A3	B3	C3	D3
BCCH carrier	f_1	f_2	f_3	f_4	f_5	f_6	f_7	f_8	f_9	f_{10}	f_{11}	f_{12}
TCH carrier	f_{13}	f_{14}	f_{15}	f_{16}	f_{17}	f_{18}	f_{19}	f_{20}	f_{21}	f_{22}	f_{23}	f_{24}
Lower band												
Frequency group	A1'	B1'	C1'	D1'	A2'	B2'	C2'	D2'	A3'	B3'	C3'	D3'
BCCH carrier	f_{13}	f_{14}	f_{15}	f_{16}	f_{17}	f_{18}	f_{19}	f_{20}	f_{21}	f_{22}	f_{21}	f_{24}
TCH carrier	f_1	f_2	f_3	f_4	f_5	f_6	f_7	f_8	f_9	f_{10}	f_{11}	f_{12}

As has been said, the traffic density across the network is rarely uniform: it is more common that cells of different sizes are used in different parts of a network (through cell splitting). This will impose difficulties in radio planning as the frequency reuse distance will be different for different cell sizes. A buffer zone has to be created, between the cells with different cell radii, in order to avoid severe co-channel interference. If the spectrum allocation is limited and additional buffer frequency sets are not available, a degradation in the network quality of service is to be expected.

In addition to the assignment of frequency group to a cell, it is also necessary to assign a base station identity code in association with the frequency group. This will eliminate the possibility of incorrect cell identification and will allow the evolution to future cell architectures. The base station identity code (BSIC) is a six bit colour code, consisting of a three bit network colour code, and a three bit base station colour code. The principle for allocation of the base station colour codes is similar to frequency assignment, but at a cluster level rather than a cell level.

5.7   Base site implementation considerations

As part of the radio planning process, it is necessary to consider a number of base site implementation issues. These include the selection of antenna, the adoption of diversity reception, the use of frequency hopping and the method of minimising time dispersion impairment.

5.7.1 Antenna beamwidth

For urban environments, it is common to use sectorised antennas. For a narrow beamwidth sector antenna, the gain at the bore site is higher than for a wide beamwidth sector antenna. This has a number of advantages, in terms of better building penetration, and a faster signal roll-off at the sector boundary relative to the bore sight. This will generally help to maintain a better carrier-to-interference ratio for the network. The downside is that at the sector edge, the signal level could be lower than the design level, if a cell-radius based on an omni-directional base site is used as a basis. For this reason, the inter-site distance has to be reduced in order to compensate for the coverage gap in between sectors.

For rural areas, if directional antennas are used, the requirement for interference protection is less stringent. A wide beamwidth directional antenna (compared to urban area narrow beamwidth antennas) can be used. This will generally improve the coverage of the base station. In practice, there is very little change in the carrier-to-interference ratio in a network with 85°, 105° or 120° antennas.

It should be noted that the antenna mounting structure could significantly distort the free space measured radiation pattern. Sufficient margins should be built into the radio network design in order to compensate for the imperfection.

5.7.2 Diversity reception

Horizontal spatial diversity is frequently employed in cellular systems for reception at the base site. Diversity reception is used to combat the fast fading caused by multipath propagation. The improvement varies with different radio environments. The basic principle for achieving improvement is to exploit the decorrelation of the signal path when the diversity reception antennas are spaced sufficiently far apart. In open environments, the diversity gain is generally very low, but in dense urban or urban environments, the diversity gain is often significant.

In real life implementations, horizontal spatial diversity reception is simple to install for a sectored base site, but it will be difficult to achieve the same gain in certain azimuthal orientations, for an omni-directional base site. It should also be noted that horizontal diversity is normally preferred compared to vertical diversity, as the spatial decorrelation onset for the former occurs sooner than for the latter.

5.7.3 Downtilt

Downtilting the base station antenna is an effective technique for controlling the radiation pattern and hence the footprint of a base site. This is commonly used to control co-channel interference either by pointing the first null of the base site antenna to the co-channel cell, or by pointing the upper 3 dB point of the antenna radiation pattern to the cell edge. The amount of downtilt is dependent on the height of the building and the cell radius.

In practice, three dimensional antenna radiation patterns are difficult to obtain and the effect of antenna downtilt could not always be certain in a real environment. Downtilting an antenna has an equivalent effect of widening the beamwidth of an antenna. To avoid this, there are now electrically tilted antennas available which have either tuneable tilt angles, or pre-determined tilt angles. Mechanical tilt brackets, however, generally allow for much greater flexibility in the adjustment of downtilt angles.

5.7.4 Slow frequency hopping

This is a special feature of GSM that aims to improve the radio performance in both noise and interference limited conditions. This has been briefly discussed in Section 5.1.3, and slow frequency hopping can potentially be used to enable a tighter frequency reuse. Unlike spatial diversity, that takes advantage of the decorrelation of the signal in the spatial domain, slow frequency hopping exploits the decorrelation of the signal in the frequency domain. This feature is available for both the up-link and the down-link, but is optionally implemented by operators.

Frequency hopping has two benefits. In noise limited systems, frequency hopping helps to average out the effects of fast fading and is particularly useful at low signal levels. The other benefit is to provide interference diversity, obtained when hopping between time slots that have different interference levels. This feature is especially useful for slow moving mobile stations when the interleaving and channel coding capability is inadequate to deal with the channel errors.

In a GSM system no frequency hopping is allowed on the broadcast control channel time slot and hence the broadcast control channel carrier. In other words, omni-directional base sites with only a single transceiver do not permit the implementation of slow frequency hopping. When using a 4 x 3 reuse pattern, and a 24 RF channel spectrum allocation, two carriers will be available per base transceiver station. However, because of the frequency hopping restriction on the broadcast control channel time slot, the traffic channel time slots on the broadcast control channel carrier will not be able to support synthesizer hopping. For this reason, an overall improvement in the quality of service cannot be guaranteed. However, with a 36 RF channel spectrum allocation, three carriers can participate in frequency hopping. This provides a minimum hopping bandwidth of 4.8 MHz, and a more significant hopping gain can be achieved.

5.7.5 Time dispersion

For digital transmission, time dispersion is a potential problem. For GSM, the time dispersion is controlled by the implementation of an equaliser with an equalisation window of 14.6 microseconds. When this is exceeded, degradation is to be expected, dependent on the relative amount of energy between the dominant signal path and the reflected signal path. From the radio planning point of view, the problem can be resolved, by either positioning the base sites close to the reflector, or by implementing more base sites in order to ensure that handover can occur when the quality of the call degrades.

5.8 Handover design

A cellular system is inherently self-healing and forgiving to system design errors. This is primarily due to the capability of handing-over a call from base site to base site when a mobile station is confronted with adverse conditions. For a GSM system, handover can be triggered either by the signal falling below a threshold level, or by detecting a larger power budget from an adjacent base site. It can also be triggered by a poor signal quality, or an excessive timing advance (too far from a base site). Many variations of the handover algorithm are possible and they are the subject of proprietary implementation by individual equipment vendors, for example, see Chapter 8.

In an urban environment, significant in-building penetration and high user capacity are required. These are commonly translated into the requirement of a high base site density. With the resultant significant overlap of coverage areas, the dominant server changes rapidly from location to location; inevitably repeated handover requests will be generated. Repeated handover will lead to an increase in the switching load in the base station controller (BSC) and the mobile-services switching centre for intra-BSC handover, and inter-BSC handover, respectively.

A possible method of reducing the amount of handover is to ensure that there exists a large signal differential between two neighbouring cells. This can be achieved by adopting narrow beamwidth antennas such as a 60° beamwidth antenna. Typically, with 90° beamwidth antennas, the signal level at the cell edge is 5 dB below the peak value, while the corresponding signal level for 60° beamwidth antennas is 10 dB. However, the disadvantage is that the signal level at the cell boundary could be significantly lower in the latter case. This could lead to poor in-building penetration, which results in poor coverage, even within close proximity to a base site. Furthermore, poor carrier to interference ratio performance could also be experienced at the cell boundary due to a lower carrier level. Another method to reduce the number of repeated handovers is to impose a longer delay between handovers. Clearly the disadvantage is that this will increase the drop-call rate, especially where a rapid response to the changes in the environment is frequently required.

The use of path loss as a handover criterion will ensure that a mobile station is always connected to a base site with the minimum path loss. This ensures that lower mobile transmit powers are used, and hence minimises the up-link interference. For a network with equal effective radiated power (ERP) from all base sites, the cell sizes will be uniform. In this case, the use of path loss as a handover criterion is valid and this is equivalent to the relative signal level handover criterion. However, in practice, due to physical limitations, or the requirement of tailoring cell sizes using different transmit powers, non-uniform base site transmit power could occur. To this end, the use of the path loss as the handover criterion could lead to adverse effects if the effective radiated power from two base sites is significantly different. Specifically, if we assume that the down-link interference level is similar for two base sites, handover from a base site with a higher effective radiated power to one with a lower effective radiated power could result in the mobile station handing over to a cell with a poor carrier-to-interference ratio. This is schematically illustrated in Figure 5.3.

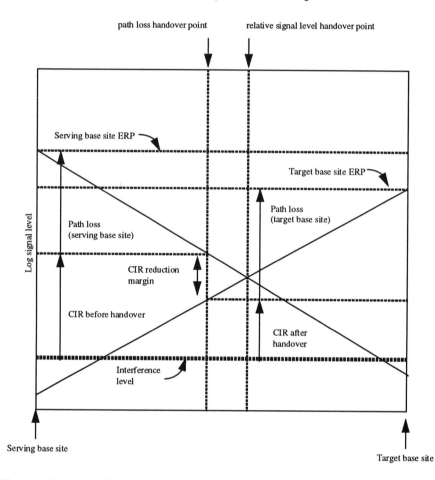

Figure 5.3   *Log distance of travel from serving base site to target base site*

In the diagram, one notes that the ERP of the serving base site is higher than for the target base site, e.g. due to practical implementation. To keep the explanation simple, the interference level is assumed to be uniform and the handover margin (also known as hysteresis) is ignored. As the mobile station travels from the serving base site to the target base site, the signal level for the serving base site gradually decreases as that for the target base site increases. Also in Figure 5.3, we see that the path loss is the difference between the effective radiated power and the measured signal level for the respective base site. When the path loss for the target base site is larger than for the serving base site, a handover will occur according to the power budget handover algorithm. However, as the target base site has a lower effective radiated power, the actual signal level from the target base site after handover is still lower than from the (previous) serving base site. This effectively leads to a decrease in the carrier-to-interference ratio. Nevertheless, if the difference in the effective radiated power between two cells is not significant, i.e. the degradation of carrier-to-interference ratio does not affect the speech quality significantly, the handover criterion could be acceptable. If the carrier-to-interference ratio degradation is significant, the speech quality will inevitably be compromised. However, in an urban environment the base sites would be densely packed, with fairly uniform effective radiated power throughout. In this situation, the use of the path loss criterion for handover becomes a valid approach.

5.9 Parameter planning

The number of parameters residing in the base station system database for a GSM network clearly far exceeds that of any analogue network. These parameters control the exact algorithm for network access, power control, handover, location updating / paging and transmissions, as well as the utilisation of other radio features of the system. They also determine the exact configuration of the radio network. The implementation of these control parameters is vendor dependent. Default settings for some of these parameters are supplied by the vendors, but they are not necessarily optimal. As part of the radio planning process, the values for all the parameters have to be determined and eventually optimised once the network is operational. This could be particularly complex for a multi-vendor network on the base station system. Frequently, manufacturers have a particular interpretation of a single GSM Technical Specification, and the handover algorithms are always proprietary.

5.10 Conclusion

In this chapter the salient points of the radio planning process have been highlighted. An insight into the actual implementation of a radio plan has also been provided. It is important to realise that spectrum allocation is of fundamental importance to the efficiency of a radio network. In addition, the user capacity and the quality of service are also critical inputs. Inevitably the investment in the network infrastructure will ultimately impose limitations on the radio coverage provided by an operator. This coverage is frequently perceived by users as one of the most important measures for comparing the merit of competing networks in a particular country.

For further reading

1. GSM Technical Specification 03.30

Chapter 6

Cellular architectures and signalling

Alastair N Brydon

Introduction

Cellular systems are a spectrally efficient means of providing wide area mobility to a large number of users, and are the established means of offering mobile communications to the mass market.

This chapter introduces key principles of cellular systems, with particular emphasis on the network aspects, and presents the state of the art of cellular network design by describing the architecture and signalling of the GSM system.

Basic principles of cellular systems are introduced, including the features needed to support mobility. The GSM network architecture, the roles of the major network components, interfaces, and protocols are also outlined. To appreciate fully the operation of the system it is necessary to understand how these various elements work together. The chapter concludes by describing typical GSM network operations, such as call set-up and location update.

6.1 Principles of cellular systems

The fundamental principle of cellular systems is that a limited radio bandwidth has the potential to support a large number of users by means of frequency reuse. Each radio frequency is used at a number of sites within the system, sufficiently far apart to avoid interference, as illustrated in Figure 6.1[1].

The support of communications in a cellular system necessitates a number of features over and above those found in a conventional fixed network. As users move around the cellular coverage area there is a need to find and communicate with them as and when required, e.g. to receive incoming calls. This is achieved by

---

1   It should be noted that this representation is highly idealised, in that the boundary between cells is not at all distinct in practice, and in general there is a diffuse region of overlap between cells.

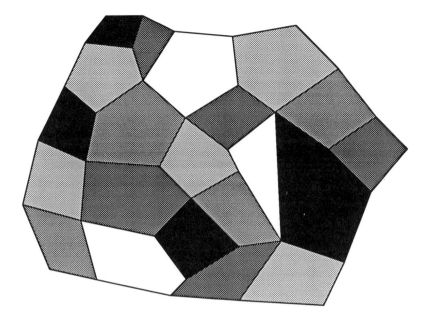

Figure 6.1   Idealised cellular radio coverage

a combination of the *paging* and *location updating* procedures. There is a further need to maintain reliable calls as users move from one cell to another, and this is enabled by the *handover* process. Finally, the provision of many services to cellular users is impacted by their movement, and this has a profound influence on the architecture of cellular networks.

6.1.1 Paging

This a process of broadcasting a message to alert a specific mobile to take some action, e.g. if there is an incoming call to be received. If a system does not know the precise cell in which a mobile is located it must undertake paging in a number of cells. In the extreme, paging might be undertaken throughout the entire coverage area of a cellular system every time a mobile is to be alerted. However, clearly this is very inefficient, and the problem is addressed by means of location updating.

6.1.2 Location updating

This is used to reduce the area over which paging must be undertaken in a cellular system. To do this, the cellular coverage area is sub-divided into a number of *location areas*. All cells repeatedly broadcast the location area in which they lie, and each time a mobile observes that it has entered a new location area it informs the network by performing a location update. The network keeps a record of the last known location of every mobile for which it is responsible.

This enables the network to perform paging over a much smaller area than would otherwise be necessary. In the extreme, it is conceivable that every cell could be a location area in its own right. The system would then know the precise cell in which every mobile is located, and need only undertake paging in that cell. However, this would generate an unacceptable level of location update signalling and processing load in the network. In practice a compromise is adopted, and location areas are defined as groups of cells.

Various enhancements can be added to the basic location update procedure, to improve further the efficiency and manageability of a network. For example, the network may instruct mobiles to *detach* from the network when switched off. The detach procedure can be viewed as the inverse of location updating, denoting that a particular mobile within a location area is *not* able to receive calls, thereby avoiding call set up attempts. In order to receive subsequent calls, the mobile must perform a location update, to *attach* to the network, when switched back on.

To maintain an accurate view of the status of mobiles within a coverage area, the network may request mobiles to perform location updates periodically, thereby confirming that they are still switched on and able to receive calls. If a mobile does not perform a location update within the expected time, it is assumed to have left the network and is implicitly detached. Periodic location updating has the added benefit of providing a recovery mechanism in the unlikely event that a network loses its location data.

6.1.3 Handover

This is the means of maintaining a call when a user moves outside the radio coverage of the serving cell; the call must then switch to a cell which does provide coverage. This must happen automatically, without loss of the call, and preferably without the user realising it has happened.

Handover is a complex process (see Chapter 8), not least because it requires the cellular system to account for the vagaries of the radio environment. It also entails close cooperation and synchronisation of events between the mobile terminal and the network. In particular, there is a need to route the on-going call to the new cell before handover can be effected, and to maintain the old connection until the handover is known to have succeeded. Handover is, by its nature, a time critical process, requiring action to be taken before an existing radio link degrades to such an extent that a call is lost. Furthermore, any interruption to the flow of the call must be kept to an absolute minimum. Handover imposes significant overheads on any cellular network.

6.1.4 Service provision

A cellular network differs in a number of respects from fixed network service provision, particularly with the proliferation of ever more complex 'network' services, over and above simple telephony, e.g. call forwarding, call barring, calling line identity, etc.

The finite radio resource available to a cellular system imposes obvious restrictions on the data rates achievable, but there are impacts on the provision of various other services, caused not by the radio interface, but by the mobility of the users. These are generally a consequence of the distribution and movement of information around the network as users move. For example, a subscriber's service profile, e.g. including numbers for call forwarding services, is likely to be held at a fixed point in his home network. The subscriber may himself move around within the network, and even make use of other networks, if this is permitted. The status of the subscriber at any given time, such as whether he is engaged in a call, is known only by the local network. Hence the provision of some services, e.g. call forwarding on subscriber busy, requires a combination of subscription information (here the forwarding number), and local information (here that the subscriber is engaged). This has a profound influence on the architecture of cellular networks, leading to distribution of service control between a fixed point in the home network and a point local to the subscriber.

6.2 Global system for mobile communications

GSM represents the state of the art of cellular network design. The system has been (and continues to be) specified by the European Telecommunications Standards Institute (ETSI), and evolves new features and services at a pace. Thus in 1995 there are some 125 operators and over 20 million subscribers worldwide.

GSM offers a broad range of services to the mass market, and enables subscribers to roam internationally between networks belonging to different operators. To achieve this, the GSM specification defines in great detail the use of the radio interface, enabling mobile equipment to communicate with any GSM network. It also defines the major network interfaces, enabling networks to exchange information related to individual subscribers. Some of the more significant features of GSM are now outlined.

6.2.1 Digital radio interface

Unlike previous cellular systems, GSM adopts a digital radio interface, enabling it to carry encrypted and error protected speech and data.

6.2.2 Subscriber identity module

A GSM user is not tied to specific mobile equipment. Users' identities and other subscriber-specific information are carried by a smart SIM card, which can be transferred between mobiles, any one of which can then behave as though it is dedicated to that subscriber. Once a subscriber has inserted his SIM into a mobile, all his incoming calls will be directed to that mobile, and all his outgoing calls will be charged to his account. Also, supplementary services will be configured according to the subscriber's preferences, e.g. call forwarding options, call barring options.

6.2.3 Integrated services

Early cellular systems focused primarily on the provision of telephony, and while they can often be adapted to support other services, this may be cumbersome, e.g. with the addition of suitable modems, a speech channel might be used to carry data or fax. Conversely, GSM offers a comprehensive range of data, messaging, and supplementary services, as well as standard telephony. GSM was designed to be the mobile part of ISDN, and offers many equivalent services. These are complemented by several GSM-specific services, including a two-way, acknowledged messaging service, SMS (described in Chapter 12).

The specification of GSM has been undertaken in phases, enabling the early introduction of basic services, followed by subsequent enhancements as developments were completed. GSM phase 1 was finalised in 1990, and provided support for telephony and simple data services. Phase 2 followed in 1994, and brought with it many improvements to the underlying operation of the system. By the end of 1996 most GSM operators will be running GSM phase 2. From this platform a range of advanced services are now being developed, known as phase 2+. These include PMR services, high rate data, and packet radio. Tables 6.1 and 6.2 provide a breakdown of the current and planned user services of GSM.

6.2.4 Roaming

First generation cellular systems are usually national or regional standards, e.g. AMPS in the US, TACS in the UK, NMT in Scandinavia. This has restricted the market for each system, and has tended to limit the scope for users to take their mobiles outside their home networks. GSM addressed this problem by standardising the radio and network interfaces, imposing rigorous type approval of mobile equipment, and requiring GSM operators to establish technical and commercial roaming agreements. As a result, subscribers can use the GSM system when abroad provided, however, their home network has a roaming agreement with the local GSM operator. Such users will have access to a core set of GSM services, but will receive bills from their home network for any network usage.

This takes place automatically, requiring little active involvement by the user. Callers to GSM subscribers need have no knowledge of their whereabouts, and indeed for privacy reasons are given no indication to that effect.

Table 6.1   Existing GSM services available in GSM phase 2

TELESERVICES (i.e. applications of underlying bearer services)
Telephony
Emergency Calls for SIM-less mobiles
Short Message Service Point to Point Mobile Terminated (SMS/MT)
Short Message Service Point to Point Mobile Originated (SMS/MO)
Short Message Service Cell Broadcast (SMS/CB)
Automatic Fax Group 3
Alternate Telephony/Fax Group 3
BEARER SERVICES (i.e. underlying communication capabilities)
Full-rate speech (13 kbps coding)
Half-rate speech (7 kbps coding)
Circuit switched data (asynchronous) at 300,1200,1200/75,2400,4800,9600 bps
Circuit switched data (synchronous) at 1200,2400,4800,9600 bps
PAD access (asynchronous) at 300,1200,1200/75,2400,4800,9600 bps
Packet network access (synchronous) at 1200,2400,4800,9600 bps
Alternate speech and data (asynchronous) at 300,1200,2400,4800,9600,1200/75 bps
Speech followed by data (asynchronous) at 300,1200,2400,4800,9600,1200/75 bps
SUPPLEMENTARY SERVICES (i.e. enhancements to basic communication facilities)
Call forwarding unconditional (CFU)
Call forwarding on mobile subscriber not reachable (CFNRc)
Call forwarding on mobile subscriber busy (CFB)
Call forwarding on no reply (CFNRy)
Barring of all outgoing calls (BAOC)
Barring of all outgoing international calls (BOIC)
Barring of all outgoing international calls except to home PLMN country (BOIC-exHC)
Barring of all incoming calls (BAIC)
Barring of incoming calls when roaming outside home PLMN country (BIC-roam)
Calling Line Identity Presentation (CLIP)
Calling Line Identity Restriction (CLIR)
Connected Line Identity Presentation (CoLP)
Connected Line Identity Restriction (CoLR)
Call Waiting (CW)
Call Hold (HOLD)
Call Transfer (CT)
Multi-Party service (MPTY)
Closed User Group (CUG)
Advice of Charge - Information (AoC-I)
Advice of Charge - Charging (AoC-C)
Unstructured Supplementary Service Data (USSD)

Table 6.2   GSM services under development in GSM phase 2+

Call deflection
Call Forward Enhancements
Completion of Calls to Busy Subscriber (CCBS) i.e. Ring Back When Free
Completion of Calls to subscriber when No Reply (CCNRy)
Completion of Calls when subscriber Not Reachable (CCNRc)
Compression of user data
DECT access to GSM networks
Direct Subscriber Access and Restriction (DSAR)
Enhanced full rate speech
Explicit Call Transfer
Enhanced Multi-level Precedence and Pre-emption Service - for PMR applications
Facsimile enhancements
General Packet Radio Service (GPRS) i.e. packet switched radio interface
High speed circuit switched data i.e. data rates > 9.6 kbps e.g. 64 kbps
Location services i.e. determining mobile position
Malicious Call Identification (MCID)
Mobile Access Hunting (MAH)
New Barring Services
Packet data on GSM signalling channels
Payphone services
Premium rate services
Proactive SIM
Support of operator specific services when roaming (CAMEL)
Support of optimal routeing
Support of Private Numbering Plan (SPNP)
Voice Broadcast Service (VBS) - for PMR applications
Voice Group Call Service (VGCS) - for PMR applications

6.3 GSM network architecture

In many ways GSM follows the principles of Intelligent Networks (IN), by separating control and service provision aspects from the underlying switching fabric. Figure 6.2 shows the GSM network architecture, identifying each of the network elements and the interfaces between them. The following outline the main functions of each element.

6.3.1 The mobile station

This is the physical equipment used by a GSM subscriber. It comprises two distinct parts: the SIM and the Mobile Equipment (ME) part.

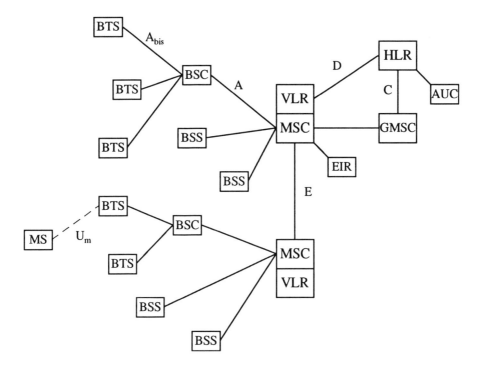

Figure 6.2   GSM network architecture

6.3.2 *The subscriber identity module*

As described, this is a smart, card which carries subscriber-specific information. The SIM also acts in security and confidentiality procedures, which guarantee the authenticity of a particular subscriber, and maintain the privacy of information carried across the radio interface.

The primary function of the SIM is to identify the current user of an MS. To this end, the SIM permanently stores an International Mobile Subscriber Identity (IMSI), which uniquely identifies a specific GSM network, and within that network, uniquely identifies a specific subscriber. The IMSI is used by the SIM during call set-up and mobility procedures, such as location update.

The SIM also carries a personal authentication key (K_i), and participates in GSM security procedures. K_i is used to prove to a GSM network that a given SIM is valid, as outlined below. The SIM also participates in the generation of cipher keys (K_c), which are used to encrypt the radio interface.

The SIM semi-permanently stores system information used during mobility and other procedures, e.g. the location area in which the SIM was most recently used, the current cipher key, and a list of networks where roaming is not permitted for this subscriber.

The final area of activity of the SIM is the storage of a small amount of user information, such as short messages and advice of charge details.

6.3.3 The mobile equipment

This contains the radio and signal processing functions needed to access the GSM network, and provides the Man-Machine Interface (MMI), which enables a user to make use of the GSM services. Depending on the ME's application it may have interfaces to external terminal equipment, such as PCs or fax machines, and may even be integrated with such equipment.

In addition to basic modulating, demodulating and transceiving functions, the ME supports a variety of features, essential to the operation of the GSM radio interface, and can be used to improve performance, e.g. timing advance, power control, discontinuous transmission, and slow frequency hopping. These features are described below.

The MMI comprises the display, keypad, and acoustic aspects of the ME. These are particularly important elements of GSM, having a major impact on users' perception of the overall system. They are becoming more important as mobile systems become increasingly sophisticated, and require simple, intuitive means of service invocation. The acoustic design must take into account the relatively long delay in GSM speech coding (about 100 ms), which can lead to perceptible echoes in networks to which GSM is connected.

6.3.4 The base station sub-system

This is responsible for providing the access between GSM mobiles and the GSM core network, and undertakes the management of the radio resource. A BSS comprises a Base Station Controller (BSC), Base Transceiver Stations (BTSs), and Transcoder and Rate Adaptation Units (TRAUs).

6.3.5 The base transceiver station

This provides the GSM radio coverage within a cell. It comprises radio transmitting and receiving equipment and associated signal processing, and is in effect a sophisticated radio modem. The BTS complements the radio features of the ME and supports power control, discontinuous transmission, and slow frequency hopping.

6.3.6 *The base station controller*

This is parent to a number of BTSs. In traffic terms, it acts as a concentrator, and provides local switching to effect handover between BTSs to which it is connected. It is also responsible for the management of the radio channel, and takes control of a variety of radio-related procedures, including handover, to ensure that reliable communications are maintained. Thus, while the BTS and/or ME are responsible for the execution of radio related procedures, such as timing advance, slow frequency hopping, and discontinuous transmission, it is the BSC which actually controls these features.

A major function of the BSC is the management of handovers. Handovers within a BSC area are generally undertaken wholly by that BSC. Handovers between BTSs belonging to different BSCs, however, can involve MSCs, but are still managed by the original serving BSC.

The BSC controls the transmission of various system information over the radio interface, including Location Area Codes (LAC), signalling channel configuration details, and information on neighbouring cells. It is also responsible for the scheduling of paging messages as incoming calls arrive.

6.3.7 *The transcoder and rate adaptation unit*

This is responsible for converting services between their radio interface format and their fixed network format. Thus, for example, the TRAU is responsible for transcoding between GSM encoded speech, at 13 kbps, and fixed network encoded speech, at 64 kbps. Similarly it performs rate adaptation of the GSM data services.

While the TRAU is nominally part of the BSS, in practice it is generally located at MSC sites. This enables network operators to benefit from the lower rate coding of services between MSCs and BSCs, and the consequent savings in transmission costs.

6.3.8 *The mobile-services switching centre*

This is essentially an ISDN switch, with (significantly) enhanced processing capability to cater for the special needs of GSM. An MSC will parent a small number of BSCs.

The primary responsibility of the MSC is call handling for the mobile subscribers within its area. This includes setting up and clearing down calls, the generation of charging records, and the execution of some supplementary services (in cooperation with an associated VLR). The MSC inevitably becomes involved in various mobility procedures.

The MSC controls the allocation of terrestrial resources to individual calls, and takes part in inter-BSC handover, where it provides a traffic switching point between the BSCs.

MSCs have Interworking Functions (IWFs) associated with them to suit particular network and service interconnection requirements, e.g. modems to support data communication across PSTNs, fax adapters to support fax group 3, interworking to data networks and ISDN, and various protocol conversion. MSCs also include *echo cancellers* to combat the threat of echo on speech circuits, caused by the inherent delay of GSM speech coding.

Whilst they are identified as distinct entities within the network, the Mobile-services Switching Centre (MSC) and Visitor Location Register (VLR) are usually implemented as a single MSC/VLR unit.

6.3.9 The visitor location register

This is an intelligent database and service control entity. It stores (on a temporary basis) the information needed to handle calls set up, or received by MSs, within its area of responsibility, and executes certain services (primarily those associated with outgoing calls). It is also heavily involved in the mobility and security management procedures of GSM.

The data carried by a VLR includes the International Mobile Subscriber Identities (IMSIs), current Location Area Codes (LACs), authentication and encryption details, and supplementary service entitlements for each subscriber currently registered with it. The VLR also keeps a local record of various service information related to these subscribers, as provided by their respective Home Location Registers (HLRs), and acts on this to execute certain services when required e.g. barring certain classes of outgoing calls.

The VLR participates in various mobility management procedures, for example by instigating paging to called users, and by passing location area information to other network entities. The VLR is also responsible for various security procedures, and initiates authentication and ciphering procedures as required.

6.3.10 The gateway MSC

This has additional functions over and above the MSC, and is the target for all calls bound for GSM users. It is responsible for determining the location of a called subscriber, and for routeing calls accordingly, i.e. towards the appropriate MSC/VLR.

6.3.11 The home location register

This is an intelligent database and service control function, responsible for the management of mobile subscribers' records, and control of certain services (primarily those associated with incoming calls). It carries subscription details such as the teleservices, bearer services and supplementary services that are to be made available to a subscriber, and location information which enables the routeing of incoming calls towards him, e.g. the MSC/VLR currently serving his MS. This information is accessed by reference to the subscriber's diallable number, the Mobile Subscriber ISDN (MSISDN) number, or the subscriber's system identity, the International Mobile Subscriber Identity (IMSI).

6.3.12 The authentication centre

This is an intelligent database concerned with the regulation of access to the network, ensuring that services can be used only by those who are entitled to do so, and that access is achieved in a secure way.

The principle is that the AuC and SIM hold a unique, individual key (K_i) for every subscriber, which is used by a defined authentication algorithm as the basis for calculating a response (SRES), to a random number (RAND), generated by the AuC. Only the true SIM will be able to generate the correct response, and thus gain access to the network. The AuC and SIM also independently generate a cipher key (K_c) which is used to encrypt the radio communications with a given subscriber. The AuC is generally integrated with the HLR.

6.3.13 The equipment identity register

This is a database carrying information on certain MEs, in order that network operators can recognise lost or stolen mobiles attempting to access the network. Each ME can be identified by a unique International Mobile Equipment Identity (IMEI), and this is used to detect mobiles which have been black listed, i.e. refused access, or grey listed (which might be allowed limited access, and monitored closely).

6.3.14 Administration and management systems

Various systems sit alongside the GSM network components described herein, undertaking activities such as subscriber management, and network operations and maintenance.

Subscriber management includes the control of subscriptions (services subscribed to, etc.) and issuing bills as required. Network operations enable the network operator to monitor the performance of the network, and to optimise this by modifying the network configuration. Network maintenance includes fault logging, and instigation of remedial action. These are key aspects of cellular systems, but are outside the scope of this chapter. Further information can be found in Chapter 7.

6.4 GSM interfaces and protocols

Figure 6.2 also shows each of the interfaces of the GSM network. A mixture of call-related and non-call-related, e.g. mobility management, signalling is carried across these interfaces, passing between various pairs of entities, e.g. call control between MS and MSC, mobility management between MS and VLR, radio resource management between MS and BTS, etc. This section considers the physical interfaces between the various GSM network entities, before introducing the protocols used on these interfaces.

6.4.1 The radio interface

This is entirely GSM specific, and is based on a combination of Frequency Division Multiple Access (FDMA) and Time Division Multiple Access (TDMA). The GSM spectrum allocation and its usage are as follows:

- 890-915 MHz mobile transmit
- 935-960 MHz base transmit
- 200 kHz carrier spacing
- 270.83 kbps data rate per carrier
- 0.3 GMSK modulation
- Optional Slow Frequency Hopping (SFH) at 10 hops per second
- Basic TDMA frame of 8 (156 bit) time slots per carrier
- Complex multiframe TDMA structure on top of the basic frame

A variety of additional features are incorporated:

Timing advance is needed by GSM to account for the uncertain propagation delay between a mobile and its serving base station, according to their separation. The principle is to time the transmission bursts made by a mobile such that they arrive at their target base station at just the right moment in the TDMA frame.

Power control is used to minimise the general level of radio interference in the GSM band, by reducing mobile transmit powers to the minimum level needed to achieve satisfactory communication. This has the added benefit of conserving mobile battery life.

Discontinuous transmission (DTX) is a means of minimising radio transmissions during the idle period in the natural flow of speech, thereby reducing interference and extending mobile battery life. During these quiet periods comfort noise is generated locally, instead of transporting 'silence' across the radio interface.

Slow frequency hopping (SFH) is a spread spectrum technique, whereby a GSM carrier changes frequency ten times per second, thereby avoiding possible periods of signal fade or interference.

The physical channels provided by the GSM FDMA/TDMA scheme are shared by a set of logical traffic and signalling channels. The usage of these logical channels is summarised in Table 6.3, classified as follows:

Broadcast downlink - information is transmitted by the network to all MSs within the broadcast area.

Common downlink - information is transmitted by the network to a specific MS, on a shared downlink channel, to which other MSs may be listening.

Common uplink - information is transmitted by a mobile to the network on a channel shared with other MSs. A slotted aloha protocol is used to resolve contention for this channel.

Dedicated uplink/downlink - information is passed privately between the network and the mobile on a one to one basis. The dedicated mode is adopted, for example, during an active call, or when undertaking subscriber-specific signalling e.g. location updating or short message transmission. At other times the MS is said to be in the idle state, and periodically monitors the broadcast and common downlink channels.

A derivation of the ISDN D-channel link layer protocol, LAPD$_m$ provides error protected links on the dedicated signalling channels (FACCH, SACCH, SDCCH) to carry the GSM protocols.

6.4.2 The BTS - BSC interface (A$_{bis}$)

This is based on conventional 2 Mbps fixed circuits. The standard ISDN D-channel link layer protocol, LAPD, is used to provide error protected links on which GSM-specific protocols can be run.

6.4.3 The BSC - MSC interface (A)

This is based on conventional 2 Mbps circuits using Signalling System number 7 (SS#7) common channel signalling. The Signalling Connection Control Part (SCCP) is used to carry GSM-specific protocols.

Table 6.3 GSM radio interface logical channels

Channel name	Channel type	Usage
Frequency Correction Channel (FCCH)	Broadcast downlink	Used by mobiles to achieve frequency synchronisation with the BTS
Synchronisation Channel (SCH)	Broadcast downlink	Used by mobiles to achieve time synchronisation with the BTS
Broadcast Control Channel (BCCH)	Broadcast downlink	Carries a variety of system information used by mobiles in the idle mode e.g. network identity, current location area code, information on surrounding cells
Cell Broadcast Channel (CBCH)	Common downlink	Carries cell broadcast short messages
Paging Channel (PCH)	Common downlink	Carries paging messages to alert mobiles to incoming calls or messages
Access Grant Channel (AGCH)	Common downlink	Assigns mobiles to specific channels for dedicated operation
Random Access Channel (RACH)	Common uplink	Used by a mobile to request resources for a subsequent operation e.g. to establish a call or perform a location update
Fast Associated Control Channel (FACCH)	Dedicated uplink and downlink	A high rate signalling channel, used during call establishment, subscriber authentication, and for handover commands
Slow Associated Control Channel (SACCH)	Dedicated uplink and downlink	A low rate signalling channel associated with each traffic channel, used for non-critical signalling such as radio measurement data
Stand-alone Dedicated Control Channel (SDCCH)	Dedicated uplink and downlink	A signalling channel which can be used independently of calls when signalling alone is required e.g. for location updates, Short Messages, supplementary services management
Traffic channel full (TCH/F)	Dedicated uplink and downlink	Carries full rate (13 kbps) speech, or data at 12, 6 or 3.6 kbps
Traffic channel half (TCH/H)	Dedicated uplink and downlink	Carries half rate (7 kbps) speech, or data at 6 or 3.6 kbps

6.4.4 The GSM core network interfaces

These are designated types B to F, and are based on conventional 2 Mbps circuits with SS#7 signalling. Call control signalling is carried using a standard fixed network call control protocol, e.g. ISDN User Part (ISUP), Telephony User Part (TUP), or National User Part (NUP), while mobility and other GSM signalling is carried by a GSM-specific development, known as the Mobile Application Part

(MAP). This part is transported by the SS#7 SCCP and Transaction Capability Part (TCAP).

Table 6.4 - Summary of GSM protocols

Protocol	Between	Function
Radio Interface Layer 3 - Radio Resource (RIL3-RR)	MS-BSC	Enables the MS and BSC to cooperate for the management of radio resources.
Radio Interface Layer 3 - Mobility Man (RIL3-MM)	MS-MSC	Enables the MS to communicate with the network for mobility management.
Radio Interface Layer 3 - Call Control (RIL3-CC)	MS-MSC	Enables the MS to communicate with the MSC for call control and supplementary service provision.
Radio Sub-system Management (RSM)	BTS-BSC	Used by the BSC to configure the BTS, and by the BTS to report measurements to the BSC.
Base Station System Mobile Application Part (BSSMAP)	BSC-MSC	Carries requests for initial connection establishment, and changes in connection attributes, between BSC and MSC. Also handles handovers between relay MSC and BSC.
Direct Transfer Application Part (DTAP)	BSC-MSC	A transport mechanism between MSC and BSC enabling communication between an MSC and an MS for Call Control and Mobility Management procedures (RIL3-CC and RIL3-MM).
Mobile Application Part (MAP)	Various	MAP has been developed as a Signalling System number 7 (SS#7) application protocol specifically aimed at supporting the special requirements of GSM. It has many different applications according to the pairs of entities using it. These are detailed below.
MAP	GMSC-HLR	Enables the interrogation of the HLR by the GMSC during incoming call set-up.
MAP	MSC/VLR-HLR	Primarily a Mobility Management protocol, but also serves to convey call-related information for incoming calls.
MAP	Anchor MSC-Relay MSC	Carries exchanges between two adjacent MSCs. It is concerned with all aspects of inter-MSC handover. Acts as a transportation mechanism for messages which must be relayed by an intermediate MSC between an anchor MSC and an MS.
MAP	MSC-EIR	A simple request/response protocol for IMEI checking.
MAP	VLR-VLR	Forwarding of parameters between VLRs.
MAP	MS-HLR	Supplementary service management.

In a conventional fixed network, the majority of the signalling is associated with call control, including supplementary services, but within GSM there is a need to convey information associated with mobility and radio resources, as well as call control, and a number of GSM specific protocols have been developed to address this. Table 6.4 provides a summary of the major protocols.

6.5 GSM operations

The roles of the various GSM network entities and protocols are best understood by considering their participation during operations.

6.5.1 Location update

This is instigated by an MS when it observes that it has entered a new location area[2]. It detects this by comparing the last known location area (stored on its SIM), with the location area information broadcast by the local cell. The MS then gains access to a radio channel, and requests a location update.

If the MSC/VLR serving the mobile has not changed, it can immediately authenticate the MS, and record the new location area. However, if the MS has moved to a new MSC/VLR, further actions are required. The MSC/VLR addresses a message to the MS's HLR, which may be in another network if the subscriber is roaming. This address can be derived from the MS's international mobile subscriber identity.

The HLR takes note of the new location of the mobile, and downloads to the MSC/VLR various security parameters, which can be used to authenticate the mobile and to establish ciphering on the radio channel, if required by the operator. The HLR also passes a subset of the MS's service profile to the new MSC/VLR, in order that various supplementary services can be configured as required. The HLR instructs the previous MSC/VLR to delete the record of the MS.

Figure 6.3 shows information flows associated with successful location updating.

6.5.2 Incoming calls

Incoming calls for a GSM subscriber are presented initially to a GMSC in his home network. From the Mobile Subscriber ISDN (MSISDN) number, dialled by the calling party, the GMSC is able to identify the HLR which hosts the called subscriber, and requests routeing information from that HLR.

---

2   GSM also supports the options of periodic location updates, and attach/detach, each of which can generate additional location updating.

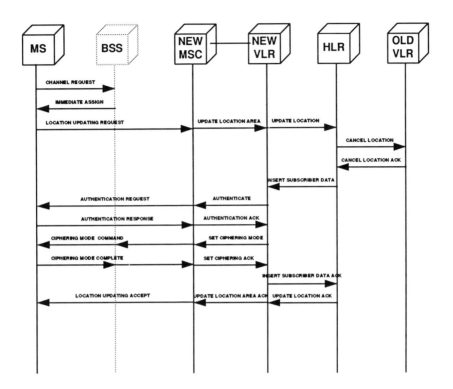

Figure 6.3   *Location update operation*

The subscriber may have active supplementary services which affect the flow of the call at this stage, e.g. call forwarding unconditional. If this is the case, the HLR provides the necessary routeing information, i.e. the forwarding number, to the GMSC, and the call is routed accordingly.

Assuming there are no such services active, the HLR must provide the GMSC with information enabling it to route the call towards the MS (which could be roaming in another network). The HLR is aware of the last known MSC/VLR for the called MS, from the most recent location update, and requests from the MSC/VLR a routeing number. The MSC/VLR temporarily allocates to the MS a Mobile Station Roaming Number (MSRN). This is a normal, diallable number, allocated from a pool of available numbers. This is forwarded to the HLR, and from there to the GMSC. The MSRN routes the call from the GMSC to the MSC/VLR (effectively the GMSC dials up a traffic link to the MSC/VLR). The MSC/VLR is then able to associate the MSRN with the specific mobile in its coverage area, and the MSRN can be returned to the pool for subsequent reuse.

At this point, the MSC/VLR attempts to establish contact with the MS by undertaking paging within the last known location area of the mobile. If, for some reason, the mobile does not respond, e.g. out of coverage or switched off, then the MSC/VLR may act with supplementary services, e.g. call forwarding on not reachable, which can then route the call to an alternative destination.

If the mobile is within coverage, and switched on, it responds to the page and requests a radio channel. The MSC/VLR authenticates the MS and establishes ciphering on the radio channel. The radio bearer can now be put in place, and the alerting signal activated on the mobile. Should there be no answer to this, then once again the MSC/VLR may take follow-on actions as defined by active supplementary services, e.g. call forwarding on no reply. If the user answers, the call can be completed successfully.

Figure 6.4 shows information flows associated with a successful incoming call.

6.5.3 Outgoing calls

Outgoing calls from a GSM user begin when the user dials a number according to the MMI of his mobile, and presses SEND. The MS firstly requests a radio channel. The local MSC/VLR authenticates the mobile, using authentication data previously obtained from the relevant HLR, and establishes a radio channel with ciphering.

If the MS has supplementary services active in the MSC/VLR, e.g. barring of certain classes of outgoing call, these may then take effect. If these have no bearing, then the call is routed by the MSC/VLR in accordance with the dialled number, and the MSC/VLR maintains charging records accordingly. If the MS is roaming, these charge records can be returned to his home network off-line, for subsequent processing.

Figure 6.5 shows information flows associated with a successful outgoing call.

6.6 Summary

This chapter has provided an introduction to the architecture and signalling of GSM, demonstrating how the system matches the basic principles of a cellular system.

GSM is a complex telecommunication system, capable of delivering a sophisticated array of services to mobile subscribers both within their home networks and beyond. The architecture is designed around the principle of separating service and mobility control aspects from the switching of traffic, and from this point of view, GSM can be regarded as one of the most advanced intelligent networks deployed in the world today. On-going developments are further enhancing the capability of the GSM system, now already established as a global leader in mobile telecommunications.

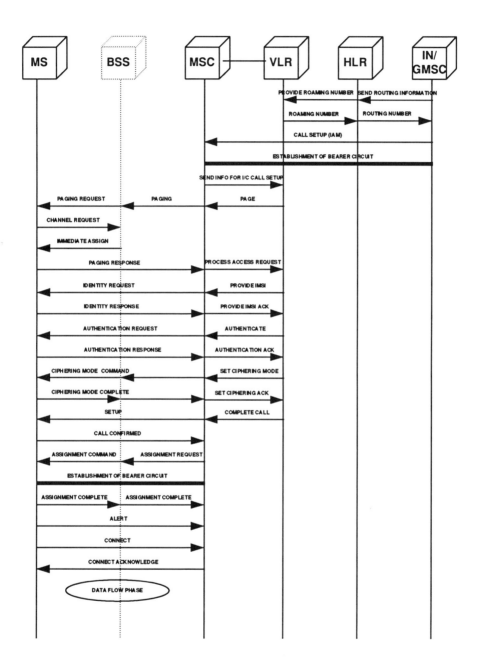

Figure 6.4   Mobile terminated call set up operation

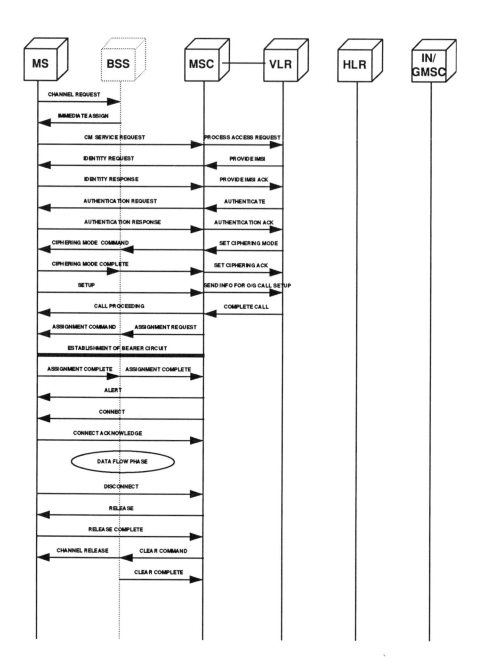

Figure 6.5   Mobile originated call set up operation

References

1   ETSI Technical Specifications for GSM, published by the European Telecommunications Standards Institute (ETSI), Sophia Antipolis, Valbonne, France

2   Mouly, M. and Pautet, M.-B., *The GSM System for Mobile Communications*, 1992, ISBN 2-9507190-0-7

Chapter 7

Cellular network management centres

John W Mahoney

Introduction

The UK cellular marketplace is the most de-regulated and buoyant telecommunications market in Europe, enjoying a significant share of the total European market. Cellnet was formed in 1983 as a joint venture between BT (60%) and Securicor.

Cellnet was originally given one of the two licenses issued by the Department of Trade and Industry to operate cellular radio services in the UK, the other being given to Vodafone. As part of that agreement both companies were assigned 300 radio channels for operation. This was called TACS (total access communications systems).

Cellnet went into service in January 1985 in London and coverage was soon extended to other major cities and towns in the UK.

Rapid growth in the market meant that additional channels were urgently needed. These were allocated in 1987, and the network became known as ETACS (extended TACS). 1994 saw the launch of the digital European system GSM, which extended the range of service available to customers, giving them the means to communicate throughout Europe and beyond. TACS, however, will continue to provide comprehensive mobile communications within the UK and operate side by side with GSM for many years to come.

The broad network architecture for both networks is shown in Figure 7.1. The strategy for ensuring effective support can be summarised as follows:

Network management
- Layered management approach in-line with the CCITT M.30 telecommunications mangement network

- A Network Management Centre (NMC) responsible for 24 hour monitoring of network elements including where possible remote diagnostics/correction
- Regional operations management units responsible for resource management (spares, test equipment etc) and field repair

Network data

- Data build and management from central data - NMC
- NMC resilience
- Creation of 'Dark NMC' for GSM
- Local terminal access provided at all BSCs

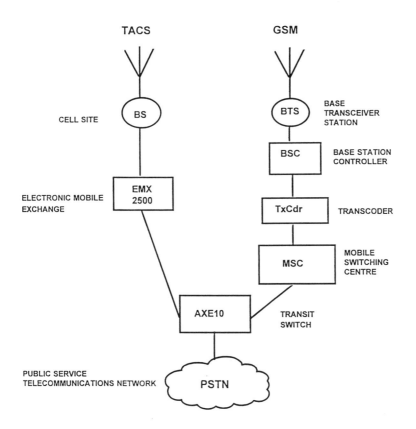

Figure 7.1   Network architecture for TACS/GSM

7.1   The network management centre

At the hub of the TACS/GSM network is a network management centre. The NMC ensures that the network delivers the best possible service to customers at all times, by identifying and clearing problems where possible, before they either affect service or are noticed by subscribers.

The rationale behind the investment in the NMC is straightforward: namely, to effectively administer a telecommunications network, a set of sophisticated network management tools is required, coupled with clearly defined processes and procedures as recommended in the Telecommunications Management Network (TMN) model, as proposed by CCITT. Cellnet, for example, has adapted and enhanced TMN to meet the exacting needs of cellular network technology.   This Cellnet network management model is illustrated in Figures 7.2(a), (b) and (c).

The model embraces performance management, network analysis, fault analysis, prioritisation and allocation, traffic profiling and network auditing as part of the generic functions.

At the business management level, Cellnet sets out its business objectives, i.e. what it wants to achieve and how to implement this; marketing and sales objectives and service management are also defined. Sales and customer services form this layer communicating objectives to the outside world and providing a service interface dealing with customer queries. The layer interfaces, on a day-to-day basis, with network management as embodied in the network management centre. This interface is physically achieved by passing network service information from the NMC in a near real time to the customer service department.

1	Business Management Level (BML)	Business planning Finance Commercial Legal
2	Service Management Level (SML)	Sales Client service Customer service
3	Network Management Level (NML)	NMC Data management
4	Element Management Level (EML)	OMUs Regional field teams
5	Element Level (EL)	Switches, cells transmission

Figure 7.2(a)   A telecommunications management network model

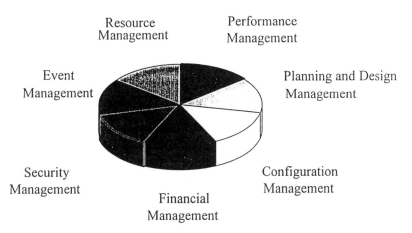

Figure 7.2(b) Generic management functions

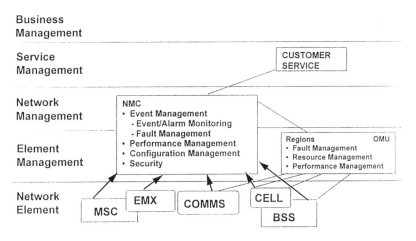

Figure 7.2(c) Allocation of TMN responsibilities for TACS/GSM

7.1.1 Network management level

This is vested in the network management centre which has overall responsibility for the quality and integrity across the whole network. Network element monitoring is carried out centrally at the NMC and hence an aspect of element management is also covered.

The functional aspects of the NMC may be outlined as follows:

- Event management
  - Event/alarm monitoring
  - Fault diagnostics
  - Fault allocation
  - Fault progression/clearance

- Performance management
  - Statistics monitoring and trend analysis

7.1.2   Element management level

This level recognises the need to manage each element as an entity. This is carried out by the provision of an Operations Management Unit (OMU) in each of Cellnet's three regions.

The OMU is the single point of contact within a boundary as defined by the region for all field operations and maintenance activities. This covers the reception and progressing of network events identified and reported by the NMC. The OMU carries out remote diagnostics, testing, fixing and general co-ordination/management of field units to assist with network repair. This includes despatching the most appropriate resource, given the nature of the event, the skill sets needed and urgency required. Once an event has been logged and agreed with the OMU, the OMU takes ownership and responsibility for the resolution of the event against agreed targets.

The OMU can be regarded as a centre of excellence within each region for diagnosing and responding to NMC reported events and ensuring regional network performance is maintained and improved. To achieve this aim, the OMU is also charged with making sure cell basesites are routinely tested for compliance against performance specification, and managing any on-site work to minimise any down time.

7.1.3   Network elements

The final level is the element level which comprises the 'building blocks' of the network, illustrated in Figure 7.3.

The systems in the NMC view the whole network (TACS and GSM) as well as managing and monitoring on a network element basis. By developing this model, one can achieve one's business objectives in a structured and formalised way, understanding the interrelation between the layers and infrastructures.

Overall, therefore, an integrated network management approach allows for an effective and efficient approach from fault detection to remote monitoring, fixing and re-configuring of network elements

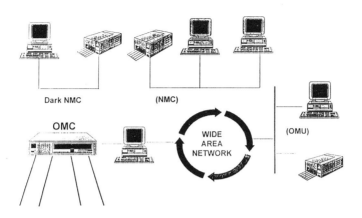

Figure 7.3   The operations management centre system network elements

From the NMC, fault detection and analysis engineers can view all major components of the network, enabling potential problems to be identified and rectified, and traffic to be effectively managed. Individual workstation positions check the network, supported by a large videowall providing a network-wide display of network data.

7.2   System features

The various systems include an Intelligent Monitoring and Control System, (IMACS), which for TACS, provides detailed information on all elements of any network, identifying faults and allocating a priority of 0 to 4 (based on the risk of service being affected). The NMC personnel can then rectify faults either from remote computer terminals or by notifying one of the field based teams via operations management units.

The report can be from any level of alarm, minor or major, and may form any level of network element, from data error to failed major hardware devices.

The IMACS system provides single line reports for individual network elements to indicate the EMX identification number, the report database reference number, time of report, priority of report, from zero to four star level of risk to service, the network element reporting an alarm condition, three columns of data to locate the reporting network element reporting an alarm condition, three columns of data to locate the reporting network device and the associated fault, supporting information such as whether a device has gone from 'in service' to out of service or an alarm has met a 'repeat' or 'duration' threshold, and finally a supporting test description of the report.

The IMACS reports may be for information only or they may give an indication that a problem is imminent, in which case preventative remedial action can be taken.

The Status and Alarm Monitoring System (SAMS) gives an immediate view of the network in graphic form, allowing engineers to identify and progress a problem from the perspective of the overall network right through to individual 'devices' following a sequence of colour indicators which appear in hierarchies on the screen, thus:

- Green indicates normal network situation
- Red alerts individual device problems

Explanatory text messages, such as 'essential maintenance work', are indicated by a neutral indication. By 'clicking' on a selected cell further detail concerning the nature of the problem is provided. SAMS provides a fast track from the NMC into customer services, where the system provides up-to-date information about the status of the network for answering customers' queries.

It is also possible to view the status of the important Inter-Switch Links (IDLs), by selecting the IDL box on the overview screen menu. IDLs are used as data communication paths for setting up calls and checking mobile validations. Their importance is marked by the fact that each IDL is duplicated.   A blue line drawn between two network nodes on the IDL display represents an IDL out of service. When a problem arises with both main and standby path, a red line is indicated.

Fluctuations in traffic demand are brought to the attention of the NMC by the Network Traffic Management System (NTMS). This system isolates and identifies areas in the network where inter node traffic is high It can then, if necessary, re-route a proportion of calls through transit switches, so reducing traffic levels in the original area and effectively 'spreading the workload' across the network.

The NTMS map displays the whole network.   'Prompts' are in the form of a colour coded line drawn between two of the network nodes. The colour coding is chosen to show the severity of the report.

Such 'prompts' indicate that an 'exception' level, or threshold, has been reached from measurements automatically computed against normally expected traffic levels.

The display can be viewed at a more detailed level down to individual switch catchment areas and associated cells. If required, a separate display of an individual measurement can be displayed for examination.

At this stage traffic thresholds can be set dynamically to match the network traffic profile. In this way NTMS can help diagnose problems, and if necessary, provide an opportunity for corrective action.

The Videowall also features a number of smaller displays such as a diary and bulletin board to alert and remind operational personnel of any planned work being carried out on the network plus more general information, such as forthcoming exhibitions and large sports events, etc., which can generate higher than normally expected traffic levels.

Also, at the NMC is a Remote Exchange Alarm Monitoring System (REAMS) which monitors power and environmental alarms at cell basesites. A further system is also provided for managing direct access connections between the network and customers, PABXs.

For GSM the network is managed via Operations Management Centres (OMCs) for each vendor's equipment. In a multi vendor network this means several OMCs for this purpose. The main facilities offered by OMCs were illustrated in Figure 7.3, and the topology is illustrated in Figure 7.4. To guard against failure of an OMC a series of 'fallback' network monitoring scenarios are provided and referred to as levels 1-3. Level 1 is normal use of the OMC. Level 2 provides for a second redundant 'fully equipped' NMC known as a dark NMC. Finally resilience is provided by direct terminal access at each BSC.

The increasing complexity and expansion of the network is paralleled by a growing number and variety of vendor supplied management systems. Left to grow unchecked this would quickly result in a diverse range of systems having bespoke man machine interfaces for network managers as a result..

To deal with this an Integrated Network Management Systems (INMS) approach is essential. Such a system provides an interface for both legacy systems and a variety of vendor specific technologies to realise a single 'object oriented' view of the 'whole' network. This approach is illustrated in Figure 7.4.

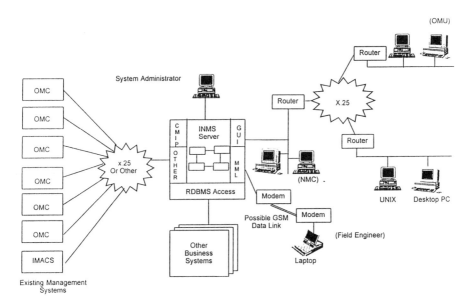

Figure 7.4 Integrated network management systems topology

By providing both 'open' and 'non-standard' front ends coupled to customised interpreters, to convert system specific data to a common format, the opportunity for a 'whole' network management system in the form of an object modelled network management system is provided.

7.3   Object modelled network management

Here the aim is to view the network as an entity comprising various 'characteristics', which are commonly grouped to form object classes. Such characteristics represent 'attributes' of the network rather than the status of an individual device or process as in the traditional case. An object could for example be defined in terms of 'service to customers' which although not a tangible element is a network attribute whose behaviour can be defined, modelled and hence managed.

In this way operational staff can view and manage the network without necessarily having complete knowledge of how each device functions. All object classes are seen at the Graphical User Interface (GUI) as near similar entities independent of equipment type and/or supplier. This approach is ideally suited for multi vendor networks and those comprising differing network architectures TACS/GSM. Object events/alerts, as seen at the GUI, 'flag' exceptions in behaviour for the attributes which characterise the object class.

7.4   Network performance management

An important aspect of the TMN model is the activity of network performance management. Here the objective is to maximise the utilisation of available capacity, whilst maintaining optimum performance in terms of call quality, dropped calls, congestion, grade of service, etc. This aim is made difficult, however, due to almost continual changes in network configuration and regular fluctuations in traffic levels. Typically the performance of the network is described in terms of Key Performance Indicators (KPI) against which targets are set and monitored. A range of commonly monitored KPIs is given in Table 7.1.

The status and trends of indicators are reported as regular management information. Where indicators fall short of targets, investigations are carried out to determine the cause of the change and corrective action implemented. The frequency of measurement and evaluation technique depends on the indicator concerned and information required.

Typically the indicators comprise three categories:

Short term - Here the source of information is usually call statistics measured in terms of traffic levels and call failure types; exception reporting is against targets set for each major network node. Since each node will have its own performance profile a system capable of modelling the individual thresholds for each node is necessary.

Medium term - This often includes regular drive testing of pre-determined routes to assess the likely call quality a typical user may experience in the area concerned. The routes are chosen to typify a particular area and using automatic call generating equipment calls are repeatedly made and their quality assessed. The resultant statistical information is captured and held on computer where it is associated with vehicle location information. Both sets of information are then at a later date displayed and examined against a geographical display so the precise location of failed calls can be identified and the network call statistics at that time examined for any correlation. Since the information is stored on computer it can be repeatedly replayed to reveal a particular call failure mechanism.

Long term - Typically switch statistics such as traffic, utilisation, grade of service on trunk routes, processor loadings, answer seize ratios and peak traffic levels are assessed over periods extending over several months. This information is most valuable in monitoring changes coinciding with new versions of network software and for major re-configurations associated with network expansion.

Table 7.1   Key performance indicators

System Availability	EMX
	CELL
	CHANNEL
Critical/Emergency cell fault	Clear within 4 hours
Major/P1 cell fault	Clear within 24 hours
Minor /P3 cell fault	Clear within 120 hours
Faults taking over 10 days to clear	Critical
	Major
	Minor
	Other
Faults Per Cell	1.8
No Fault Found	40%
Planned Outages days notice	Cat A 95.0% > = Days
	Cat B 95.0% > = Days
	Cat C 95.0% > = Days
	Cat D 95.0% > = Days
Customer Care faults	Cleared within 4 days
	Cleared within 10 days
Call Access faults	Cleared within 24 hours
	Cleared within 8 hours
	Cleared within 4 hours

Monitoring these trends implies the management of a historical database of performance measurements spanning weeks, months and even years depending on the indicator. The results of such measurements provide valuable information for network design teams and planners who need to compare predicated measurements against those actually experienced. Optimisation teams also have a use for such information to study the effects of changes in network configuration and adjustment made during optimisation activities. The information is also of great use to marketing activities to monitor the variation in traffic against a particular marketing campaign.

The main problems associated with achieving performance management include:

- The historical databases rapidly grow and require many gigabytes of disk space.
- Often the measurements generated by the network elements are in a state that require complex analysis/processing to calculate the key performance indicators.
- In many instances there are anomalies between measurements from different network equipment and different vendors.
- Continuous changes to network configuration, particularly during the network roll-out phase, need to be accurately tracked in the configuration database of the performance management system.

Sophisticated software technology is often used to overcome these problems. Metrica/NPR is an example of such technology which has been deployed to cater for both TACS and GSM. It consists of a multi tasking front end having inputs customised to receive network data from a variety of TACS and GSM element mangers including OMCs. The data from the various sources is analysed and where appropriate interpreted to provide a common measurement standard for each parameter across the whole network.

Future developments of the technology can be expected to include:

- The automatic detection and notification of when 'short term' indicators exceed target values. These conditions may be as simple as crossing over a simple threshold, or may be time or statistically variant; in order to differentiate between transients and longer term effects. The resultant reports can be expected to interface to an object orientated system and be presented as an integral feature of the GUI display.
- Use of powerful computer graphics and visualisation techniques, i.e. data mining, to help present complex arrays of data so correlations can be identified and common abnormalities easily identified.
- Geographical display of performance measurements - a problem with one network element may produce secondary problems with neighbouring elements in other network devices. For example, an abnormally high level of handover failures with one cell may be caused by congestion or a planned outage with its neighbour. Presenting this information in a geographical context can help identify secondary

problems and avoid lengthy diagnostics. Such a system is already in use in Cellnet for TACS and is a derivative of the SAMS system described earlier.

Acknowledgements

The author would like to express his thanks to the following who have contributed to this chapter:

Mark Farmer, Metrica UK Ltd
Jon Craton and Don Gibson, Information Processing Ltd

Chapter 8

Mobility handover and power control in GSM

Vernon Fernandes

Introduction

Mobility for the subscriber is the essence of a cellular system. It allows the subscriber to make and receive calls at any location in a network, provided there is radio coverage and the subscriber demands a service which has been registered. A cellular network's coverage area is made up of a series of cells, each of which can be served by one or more transceivers, whose function is to code/decode and modulate/demodulate speech or data signals. The size of the cells and the number of channels (proportional to the number of carriers available) is determined by the subscriber population distribution. The location of base station transceivers is the key to providing good rf coverage, which is important for the maintenance of an acceptable quality of service. Normally the transceivers associated with a cell are located at a base station site, but in some cases they can be distributed to improve coverage. Another variation is to have a single base station support a number of sectored cells.

Mobility management is the cornerstone of cellular philosophy. In GSM, a considerable amount of effort has been directed to providing an efficient and reliable system that allows for international and interoperator roaming. Recent developments in the standards activities have involved the study of roaming between bands (such as GSM and DCS1800) and roaming involving GSM and DECT and other mobile communications standards. There are three aspects to mobility management, namely:

- the radio aspect
- the network aspect
- the personal aspect

Since good rf links are essential to obtaining good quality, radio mobility procedures ensure that upon power up, the mobile station associates with the best base station in terms of signal strength before it enters the call handling state.

Also, during a call, the MS may be handed over from cell to cell, or even between channels in the same cell, in order to maintain a good link at all times.

In network mobility the mobile establishes its Location Area (LA) by identifying the best cell to camp on and informs the network of the Location Area Code (LAC), via a message interchange sequence.

Personal mobility, on the other hand, refers to the ability of a user to access his range (or a subset in some cases) of the subscribed services with any mobile equipment through the use of a smart card, the Subscriber Identity Module (SIM). Note that the MS is made up of the SIM and the Mobile Equipment (ME). Personal mobility can also extend to foreign networks if a roaming agreement exists with the user's home network.

8.1 The GSM network

The reference model of a GSM network, normally called a Public Land Mobile Network (PLMN), is shown in Figure 8.1. The philosophy of operation is akin to ISDN and many of the ISDN specifications have been adopted for GSM. The network entities have already been fully defined earlier in Chapters 6 and 7, for example.

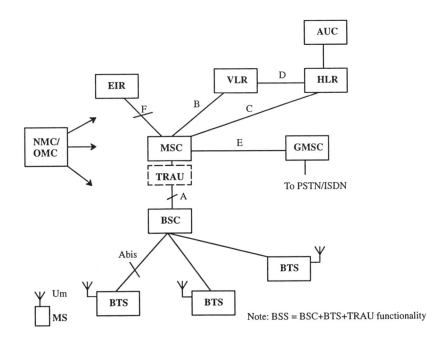

Figure 8.1   GSM PLMN reference model

For a cellular system to track a mobile requires data registers that store the location and characteristics of the mobile. In terms of a GSM network the ultimate store of such information is the HLR, which is effectively the central subscriber database. In a simple system it would be accessed every time the mobile has to communicate. This is impractical in a worldwide system where the signalling load requirements to exchange information with the HLR would be significant. Hence GSM also defines VLRs, which provide a very efficient mechanism to distribute the signalling load. VLRs retain all the necessary information for authenticating and servicing a mobile within its area of jurisdiction and are used to handle the initial signalling and routeing information pertaining to a mobile. The accessing of the HLR is therefore reduced to the identification of the VLR and the uploading of the data store in the VLR which holds the subscriber details. Once a VLR has successfully recorded the details the procedures of routeing, verification and authentication are handled by it. Meanwhile, the HLR is a fault tolerant database, which is often situated at the operator's central administration centre. It holds semipermanent and transient data.

The semipermanent data record would contain the following:

. mobile/subscriber's identity numbers
. services/capability
. restrictions

Transient data is related to mobility and is updated by means of signalling procedures with the VLR. The main elements of transient data are:

. address of the last known VLR that served the mobile
. address of the subscriber data in the VLR database
. subscriber's restrictions relating to the relevant network

The VLR is often implemented within the MSC framework, but is a separate entity. This implementation is chosen because of convenience and cost effectiveness. It is also possible to realise the VLR to be physically similar to the HLR. The VLR contains more precise location information and is updated every time there is a location update of the mobile.

A structured procedure, e.g. call set up, location update or handover, involves interaction between most of the network entities with protocols defined in the specifications. As an example, when a mobile originated call is made, the MS first accesses the BTS to request a dedicated resource on the radio interface. If the call request is accepted the BTS grants a channel and the signalling between the MS and the network can commence. During that process the MS is authenticated by checking the identity of the subscriber and the identity of the equipment against the stored information in the AUC. This process is expedited by using the VLR, which tends to be more local to a roaming subscriber. Call data is logged at the OMC along with other performance and alarm information from the network entities so that the operator always has a reliable assessment of the network.

In this chapter we consider the structured procedures relating to location update and handover, which are specially designed to handle mobility in the GSM PLMN. Apart from structured procedures, the evaluation of the radio environment for cell selection and handover by the mobile and base station is based on sophisticated real time algorithms in both entities, the key to the initiation of mobility procedures.

8.2 Radio aspects

The quality of a link is heavily dependent on getting the best radio connection in the given circumstances. As the MS moves this can be a complex process, and its management is the subject of considerable work in the specification process and design of a network. In the GSM specifications, procedures to ensure that (in the idle mode) the mobile camps on the best cell and reselects better cells have been defined in GSM 05.08 [1]. Another aspect of radio mobility is handover, which can be defined as the directed change of the radio channel between the MS and network when a dedicated link has been established. Handover helps in the maintenance of a good quality radio link and therefore a good quality of service. Before proceeding to describe radio mobility, let us review the key aspects of the GSM radio interface.

8.2.1 The GSM radio interface

GSM has adopted a TDMA structure, modulating an rf carrier to occupy a 200 kHz bandwidth about the carrier at the radio interface. In standard GSM there are 124 carriers in the 900 MHz band but this was later increased to 174 carriers by allowing an extension band. In DCS 1800 the number of carriers is 374.

The fundamental unit from which a frame structure is built up is called a bit and is identified by a Bit Number (BN). The Bit Period (BP) is 48/13 µs (~3.69 µs), and is taken as the reference unit in defining the frame structure and hierarchy.

A TDMA frame is made up of eight timeslots as·shown in Figure 8.2. The frame is repeated continuously to make up the baseband structure that is translated to the GSM/DCS frequency band by the modulation process. Each timeslot is a window which holds a group of bits known as a burst. There are five types ofburst, namely the normal burst, frequency correction burst, synchronisation burst, dummy burst and access burst. A timeslot is 156.25 BPs long, but only contains 148 bits, which constitute a burst, except in the case of the access burst, which has only 88 bits. The remaining time of 8.25 BPs (or 68.25 BPs in the case of access bursts) is known as the guard period.

Being a TDMA system, a multiframe structure is used to create a number of separate channels that can be interleaved in the timeslot stream. In the GSM TDMA system there is a 26 frame multiframe, which is used to provide 24 frames for a Traffic CHannel (TCH) and 2 frames for the Slow Associated Control

CHannel (SACCH). This is illustrated in Figure 8.3 below. In a full rate system timeslot 12 is used for the SACCH and timeslot 25 is idle.

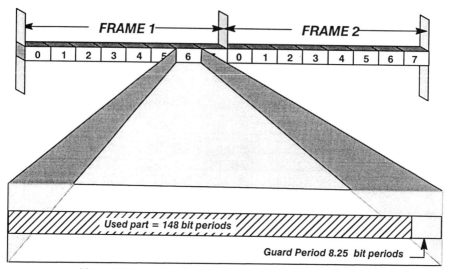

Normal, Freq. correction, synchronisation, dummy burst

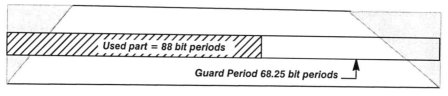

Access Burst

Figure 8.2   GSM 'bursts' and the guard period within frame structure

The GSM standard also defines a 51 frame multiframe for the remainder of the logical channel types, which are:

- Broadcast Control CHannel (BCCH): Used to provide mobiles with the Base Station Identity Code (BSIC) and channel information. Downlink only
- Frequency Correction CHannel (FCCH): Used by mobiles to track thefrequency of the base station's transmission. Downlink only.
- Synchronisation Channel (SCH): Used by the base station to transmit its BSIC and provide channel information. Downlink only

- Random Access CHannel (RACH): Used by a mobile to initially contact a base station it has 'camped on'. Uplink only.
- Paging Channel (PCH): Used by the network to initially contact a mobile. Downlink only.
- Access Grant CHannel (AGCH): Used by the base station and mobile for the purposes of signalling in a GSM structured procedure e.g. call set up. Downlink only.
- Standalone Dedicated Control CHannel (SDCCH): Used by the base station and mobile for the purposes of signalling in a GSM structured procedure e.g. call set up. Uplink and downlink.
- Fast Associated Control CHannel (FACCH): Used for signalling during a transaction. The FACCH can be associated with a TCH or an SDCCH. Uplink and downlink.

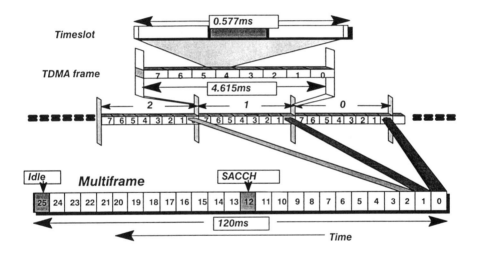

Figure 8.3   *Logical to physical mapping of the TCH/SACCH channel combination*

A number of combinations of control channels are defined in GSM 05.02 [2]. The philosophy of interleaving previously described for the TCH/SACCH subchannel is also relevant to the other control channels except that a 51 frame multiframe is used.

8.2.2 Timing advance

Normal bursts pertaining to the relevant channel type are carried in a timeslot window. As a mobile moves away from the base station, the bursts from the MS will be delayed, due to the increasing propagation delay. The uplink is timed relative to the frame structure in the downlink; hence the delay in the burst is equivalent to twice the distance between the MS and base station. The slippage experienced at the base station of an MS some 35 km away is about 230 µs, a lot more than the 8.25 BPs (30 µs) allowed by the guard band. To solve the problem, GSM has defined a timing advance command, which is transmitted by the base station, for the purposes of ordering the mobile to advance its transmission, so that the bursts fall within the prescribed window when they arrive at the base station receiver. This process is known as adaptive frame alignment and takes place continuously if a mobile has a dedicated link to the base station.

The timing advance field in the downlink command is coded with 8 bits allowing for 63 steps. Each step is one BP, or 3.69 µs. This gives a total range of 233 µs approximately. The commands relating to timing advance are very important in the assignment process during call setup and handover procedures. We return to this matter and its influence on handover, after discussing the cell selection and reselection process.

8.2.3 Cell selection/reselection

The path loss criterion parameter used to determine the best cell for selection is known as C1 in the GSM specifications [1]. It is a measure of the strength of the link between base station and mobile. Hence a base station with the best C1 is chosen by a newly arrived MS. C1 is defined using dBm signal levels as follows:

$$C1 = (A - Max(B,0))$$

where   A = RXLEV - RXLEV_ACCESS_MIN
         B = MS_TXPWR_MAX_CCH - P
         P = Maximum rf output power of the MS.

RXLEV_ACCESS_MIN = Minimum received level at the MS required for access to the system.

MS_TXPWR_MAX_CCH = Maximum TXPWR level an MS may use when accessing the system until otherwise commanded.

The MS chooses the cell with the best C1, provided it is positive.

At the boundary between cells you can have a situation in which C1a = C1b relative to adjacent cells a and b. Note that this boundary varies with different MSs, because the maximum transmission power can vary according to the type of unit. The normally rapid variations in radio conditions can therefore cause the MS to toggle between cells at the cell boundary. If the cells are in different location areas the MS will be required to go through a location update procedure each time which needs to be avoided. To prevent this from happening at the border of location areas, a further parameter, known as CELL_RESELECT_HYSTERESIS is broadcast by each cell. This is used to make the C1 value for the neighbouring cell worse than it actually is by the hysteresis value.

Cell selection at power up of the MS can be a lengthy process if the MS has to scan all frequencies in the GSM 900 or DCS 1800 bands. To mitigate this problem there is provision to store a list of frequencies obtained from the network the last time the mobile camped on the network. Since there is higher probability that some of these frequencies remain valid at subsequent switch on, the cell selection process time can be reduced by suitably ordering the frequency selection process.

For emergency or abnormal calls special allowances have to be made. In the case of emergency calls it is a requirement that the MS can operate without a SIM and any PLMN is accessible. If there is no SIM or no suitable cells of the selected PLMN or the network rejects the MS the following procedures hold:

(a) The MS monitors all carriers and camps on a cell with C1>0, which is also not barred - irrespective of PLMN.

(b) Searches for the 30 strongest GSM rf channels to determine which PLMNs are acceptable.

(c) Reselection is only performed among the cells of the 'camped on' PLMN.

8.3 Microcells

Cell sizes in a GSM system can be up to 35 km in radius and are usually serviced by a multi carrier BTS, which produces enough rf power to illuminate the cell area. The antenna is placed on a mast and is very often designed for diversity reception. In order to increase capacity with a fixed set of frequencies, cell sizes can be made smaller, enabling frequencies to be repeated more often. Alternatively cells can be sectorised, giving opportunity for more elegant repeat patterns with the net effect of increasing capacity.

An alternative method for obtaining greater capacity is to use microcells, small cells of about 300 metres radius that exist within larger macrocells. In urban areas the antennas for microcells are installed below roof tops and therefore illuminate street 'canyons'. In other words the cell's shape is found to follow the path of the roads in the vicinity of the antenna with coverage falling off drastically after turning a street corner: see Chapter 10.

The introduction of microcells in the GSM phase 2 specifications creates new system challenges. One main issue is the requirement to encourage a slow moving MS to camp on a microcell whilst fast moving mobiles would camp on to a macrocell. To satisfy this requirement a new reselection criterion C2 has been defined.

The logic of operation is as follows: The microcells in a network transmit a timer value to the MS in the BCCH downlink. This timer is used in the calculation of C2. When the timer is running, the value of C2 is artificially reduced, making the microcell appear worse, so that it is less likely to be selected and the MS would stay on the macrocell. When the timer expires the value of C2 for the microcell is boosted, and the MS is more likely to reselect the microcell. In this way a slow mobile would be encouraged to be in the same microcell coverage area for a greater period. On the other hand, a fast mobile will tend to have the timer running throughout its passage across the microcell and therefore will remain camped on to the macrocell.

Here is the mathematics:

The reselection criterion C2 is defined in GSM05.08 section 6.4 as follows:
C2 = C1+CELL_RESELECT_OFFSET-TEMPORARY OFFSET*H(PENALTY TIME)for PENALTY TIME coded 0-30
C2 = C1 - CELL_RESELECT_OFFSET for PENALTY TIME = 31

where

$$H(x) = 0 \text{ for } x < 1$$
$$= 1 \text{ for } x >= 0$$

From the above equation one can see that the TEMPORARY OFFSET value makes C2 appear worse. When the timer expires the TEMPORARY OFFSET factor is removed and C2 is boosted. The timer value can be set at the OMC such that the speed criterion can be varied depending on the anticipated traffic behaviour in the coverage area.

8.4 Network aspects

Network mobility involves the tracking of a mobile and offering the subscribed facilities to the mobile, wherever the user is located. In order to understand the system operation it is worthwhile to define first the numbering scheme and associated terminology.

8.4.1 Network area numbering

Since the mobile network is considered as an extension of the fixed network the subscriber must conform to the PSTN/ISDN numbering plan of a country. Hence every mobile subscriber will have a Mobile Station International ISDN number (MSISDN), the number one would find in the phone book. However, to improve privacy for the subscriber, GSM also defines another number, known as the International Mobile Subscriber Identification (IMSI), which is mapped to the MSISDN by the operator. The IMSI is used in the mobile related signalling transactions. The structure of the MSISDN and IMSI are given below.

To assist in the creation of the MSISDN and IMSI, and facilitate tracking of mobiles, there is a hierarchy of numbers used in a GSM PLMN, as illustrated in Figure 8.4 The definitions of these numbers are:

CC: Country Code

MCC: Mobile Country Code - Used to identify the home country of the PLMN within the IMSI

NDC: National Destination Code - Uniquely identifies a PLMN in a country for use in the MSISDN

MNC: Mobile Network Code - Identifies a GSM PLMN in a country for use in the IMSI

LAC: Location Area Code - Identifies a location area (group of cells) in a PLMN; up to 2 octets in length

LAI: Location Area Identity - Similar to the LAC but with unique international identification; made up of MCC+MNC+LAC

CI: Cell Identity - Defines a cell or sector; 2 octets maximum length

CGI: Cell Global Identity - Sent on the BCCH and provides a PLMN based cell identity; made up of LAI+CI

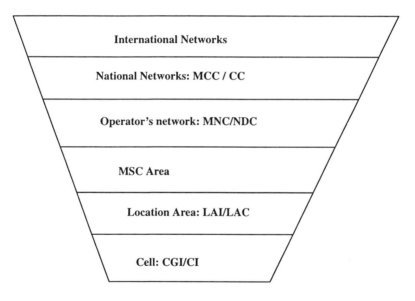

Figure 8.4   Network numbering hierarchy in the GSM system

8.4.2 Mobile related numbering

Apart from the above area or network related numbers, there are a variety of numbers related to the subscriber and the mobile equipment in use. The interactions between these numbers and the network numbers are now described.

SN: Subscriber's Number, associated with the SIM card and unique to the user.

MSISDN: Mobile Station International ISDN number; equivalent to your phone's publicised number. It consists of the CC, NDC, Home Location Register Identity (HLRID), and Subscriber Number (SN), as in Figure 8.5.

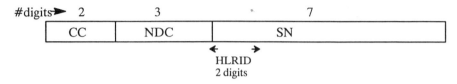

Figure 8.5   MSISDN structure

MSIN: Mobile Station Identification Number, which uniquely identifies a subscriber within a home PLMN, as shown in Figure 8.6.

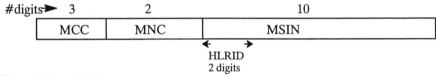

Figure 8.6   MSIN structure

IMSI: International Mobile Subscriber Identification, known only to the operator to ensure subscriber confidentiality. A unique IMSI is allocated to the MSISDN. The IMSI is used for setting up signalling connections. It consists of the MCC, PLMN code, HLRID and MSIN.

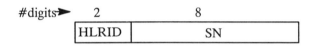

Figure 8.7   IMSI structure

TMSI: Temporary Mobile Station Identification; allocated by the VLR after authentication and used on the radio path. It gives protection from intruders listening on the signalling link; hence the TMSI is normally used in paging (though the IMSI can also be used). TMSI is more efficient, being a smaller number.

To guard against double allocating TMSIs, the number includes the current value of the system recovery timer. It also has a time stamp and the access index, which identifies the MS in the VLR. The TMSI can be reallocated at any time, but is usually changed during the location updating procedure. Re-allocation would occur when the radio link is in a ciphered mode to prevent eavesdropping. The LAI is used together with the TMSI to uniquely identify a subscriber. Both numbers are sent to the network by the mobile during the start of a location update procedure described below.

LMSI: Local Mobile Station Identification, used by the VLR database to search for subscriber details. It is the temporary address of the location containing the subscriber data.

MSRN: Mobile Station Roaming Number facilitates the routeing of a traffic channel from the gateway MSC (GMSC) to the visited MSC (VMSC). The MSRN has a maximum of 15 digits, made up from the CC, NDC and an individual number related to the ISDN telephone numbering plan, shown in Figure 8.8.

#digits➤ 2	3	= <10
CC	NDC	number

Figure 8.8   MSRN structure

HON: HandOver Number is a unique number used to establish a connection when there is an MSC to MSC handover. It provides routeing information and is generated by the target VLR during the handover process.

CKSN: Cipher Key Sequence Number is used to track the cipher key Kc. The encryption key Kc is calculated separately in the network and the MS. It therefore becomes possible that the two values are inconsistent. This potential problem is alleviated by the VLR tagging the CKSN number to the Kc locally. while sending the same number to the MS during the authentication procedure. When the MS subsequently accesses the network, the CKSN is sent back in order that the network's stored Kc can be verified.

8.4.3 System requirements for network roaming

Subscriber data, such as the IMSI, authentication key, encryption key, class of service, VLR address and directory number are held in the HLR/AC, which is normally centralised in the subscriber's home network. It can also be implemented in a distributed form. The roaming subscriber's mobile must have its location information in the HLR updated by means of the local (visited) network signalling back to the HLR the necessary data, because it is the HLR that is accessed for location information at the start of a call set up procedure. The HLR is kept updated with the mobile's location and capability details via the location update procedure.

While the HLR is able to offer location information to the network, when (say) a call is to be set up, the traffic connection within the terrestrial part of the PLMN is determined by the MSRN. The MSRN, which has ISDN numbering information for the destination point, is extracted by the HLR from the visited VLR and sent to the source MSC. This process is best illustrated with a simple example of a mobile terminated call setup.

Call set up

In Figure 8.9 the information flow relating to a mobile terminated call is given. The steps are as follows:

[1]   The caller dials the MSISDN of the subscriber to be contacted. The call is routed to the nearest GSM gateway (GMSC) in the country of origin.

[2]   The GMSC interrogates the HLR over a common channel signalling #7 mobile application part (MAP) link by sending the MSISDN to the HLR.

[3]   If the subscriber has no restrictions the HLR requests, via a C7 link, an MSRN from the VLR. At the same time the HLR sends the MSISDN, IMSI, LMSI, and optionally the MSC number, to the VLR in order to update the database. The LMSI is the address of the transient subscriber data in the VLR and enables faster access for the HLR. The VLR obtains the LAC.

[4]   The VLR sends the MSRN to the HLR.

[5]   The MSRN is transferred to the GMSC, which uses it to perform routeing to the area where the MS was roaming when the last location update was executed.

[6]   The GMSC selects a trunk to the visited MSC. If it is foreign, the PSTN and associated signalling systems required for that connection would be deployed.

[7]   The VMSC sends a page request message to the BSS.

[8]   The MS is paged in the location area over the Paging Channel (PCH).

[9]   The MS responds on the Random Access Channel (RACH).

[10] A traffic channel is set up with end to end connectivity, enabling speech or data exchange to commence. The aspects of the call setup scenario relating to authentication, ciphering and alerting are not shown in the diagram

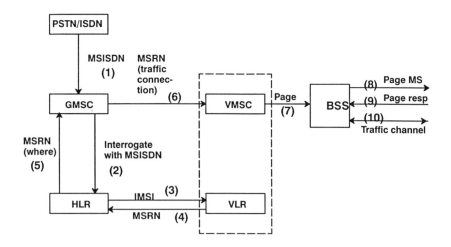

Figure 8.9   Number management in a mobile terminated call

8.5 Location updating

For a mobile station in the idle state the mechanism used to establish the location of the mobile is known as location updating. It is a structured procedure, which is initiated by the MS, and used to update the MS details in the HLR and current VLR. There are various location update procedures as determined by the following conditions:

- first location update (mobile is turned on or SIM card is inserted)
- the MS moves within the same VLR area, but to a new location
- the MS moves to a new VLR area, which is a neighbour
- the MS moves to a new VLR area, which is not a neighbour
- expiry of periodic location update timer

In all cases the parameters which are updated in the VLR are the LMSI(x), IMSI, MSISDN, HLR address, TMSI(x), LAI(x), CKSN(x) and the subscription details of the MS; the letter 'x' is used to refer to the location area in use.

8.5.1 First location update

The key steps relating to the signalling of mobility information when a SIM card is used for the first time and the MS is turned on are given in Figure 8.10 and now described:

[1] The location updating procedure starts with the mobile sending the IMSI, LAI, CKSN and its A5 capability list to the VLR. This data is normally stored in the SIM card.

[2] Recognising that no entry for the mobile exists (and that the mobile does not come from another VLR area), the VLR sets up a signalling link with the HLR with the help of the HLR code in the IMSI.   The VLR requests subscriber data from the HLR, which updates its own database with the new mobility data, VLRid and the LMSI information received from the VLR.   The LMSI gives the address location in the VLR of the MS data and is used to expedite the update of the data when the HLR replies.

[3] The HLR checks subscription status, obtains authentication parameters from the AC, establishes a backwards signalling connection to the VLR and sends all the subscriber information and its return address to the VLR. The set of numbers sent by the HLR to the VLR are the RAND, SRES and Kc. This group is known as the triple. Up to five triples can be sent at one time, to reduce the need for VLR-HLR signalling, should another location update of similar type occur again.

[4] The VLR stores the information in its database at the location addressed by the LMSI and then starts the authentication procedure.   Authentication involves the transmission of a random number (RAND) to the MS and checking of the Signed RESponse (SRES) with the locally stored version.   SRES is computed through the encryption algorithm and local encryption key in the MS and should tally with the equivalent computation that would have been performed on the network side.

[5] When authentication is successful the VLR allocates a TMSI to the MS and initiates ciphering by sending key Kc to the BSS.

[6] The A5 algorithm preferred by the network is forwarded to the MS to enable it to enter ciphered mode. The algorithm is selected by the BSS from the MS list, MSC list and its own list of supported algorithms.

[7] The TMSI is forwarded in the ciphered mode to the MS which stores it in the SIM. The SIM also updates the values of CKSN and the LAI.

[8] The mobile then sends a TMSI re-allocation complete message to the MSC via the BSS. The BSS then clears down the signalling channel and completes the procedure.

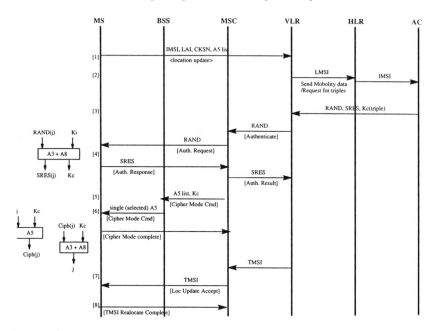

Figure 8.10   Management of mobility/security numbers during first location update

8.5.2 Location update in the same VLR area

As the MS continuously monitors the radio receive level at its current location and compares it with neighbouring locations, the MS can decide to select another cell which offers better rf conditions. If this cell is governed by the same VLR the resulting location update procedure is a simplified form of the first location update sequence and is illustrated in Figure 8.11. The main difference is that the HLR and AC are not involved in the location update procedure. The location details are altered in the VLR, without involving the HLR, and the VLR also generates a new TMSI to be forwarded to the MS in a ciphered form.

The authentication process is optional, but the cipher key K_c has to be sent to the new BSS via the MSC.

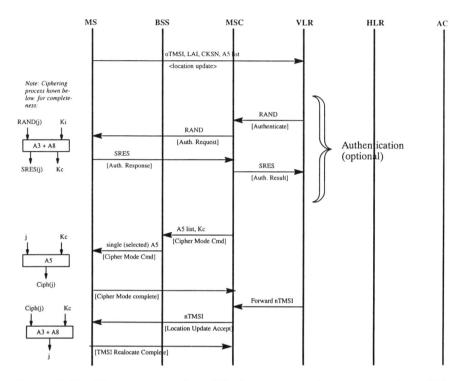

Figure 8.11   Management of mobility/security numbers during intra-VLR location update

8.5.3   Location update with a new VLR which is a neighbour

The management of mobility data when a mobile moves to a different neighbouring VLR, while in idle mode, is handled by the location update procedure illustrated in Figure 8.12. The important steps are numbered and described below:

[1]   The MS sends the old TMSI (oTMSI), LAI, CKSN and A5 list in non-ciphered mode to the VLR.

[2]   From decoding the LAI received, the new VLR detects the fact that the MS would have come from a neighbouring VLR. The new VLR creates data space for the MS, with address being a new LMSI, and subsequently sets up a signalling channel to the old VLR. It uses that channel to send the received TMSI and LMSI to the old VLR in order to obtain the IMSI and unused triples.

[3]   The new VLR establishes a link to the HLR with the help of the HLR code present in the newly received IMSI. The new VLR uses this link to send its address (VLRid) and LMSI to the HLR which in turn replies with the subscriber data, HLR return address and a set of triples.

[4]   The HLR indicates to the old VLR (via the location cancellation procedure) that it is allowed to relinquish data belonging to the mobile. The address space may be overwritten by a new subscriber.

[5]   Authentication may be optionally invoked.

[6]   Ciphering information in the form of the A5 list and Kc is sent to the BSS from the new VLR. This enhances confidentiality and gives some flexibility if the new BSS does not support a particular A5 algorithm.

[7]   The TMSI is forwarded in the ciphered mode to the MS and the old TMSI is overwritten in the SIM. At the same time the new LAI and CKSN are recorded in the SIM.

[8]   The MS acknowledges receipt and the processing of the TMSI by returning a TMSI reallocate complete message to the network.

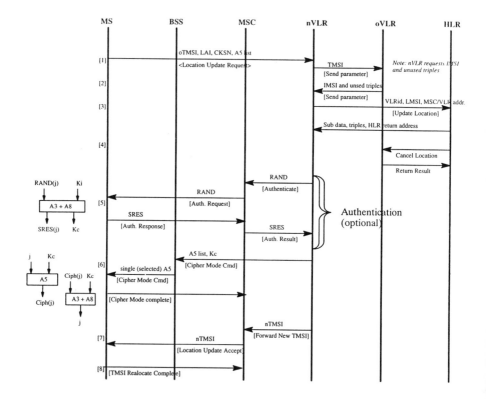

Figure 8.12   Management of mobility/security numbers during inter-VLR (neighbour) location update

8.5.4 Location update with a new VLR which is not a neighbour

The main scenario difference between the location update with a VLR, which is a neighbour, and a VLR, which is not a neighbour, is the absence of the VLR-VLR signalling link, in the latter case. As such, the IMSI and unused triples are not recoverable from the old VLR. However, the IMSI is essential if the HLR is to be addressed by the new VLR (to extract and update mobility data). Hence the new VLR immediately requests the IMSI directly from the MS. The other aspects of the procedure are very similar to the non-neighbour case and are given in Figure 8.13.

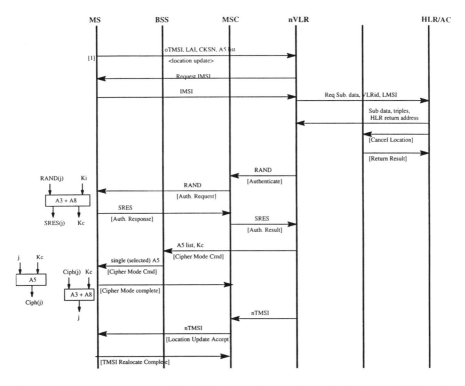

Figure 8.13   Management of mobility/security numbers during inter VLR (not neighbour) location update

8.5.5 Periodic location updating

If a network requires the MS to perform periodic location updating, a flag is sent on the BCCH. This indicates to the MS that it should perform the location update procedure periodically. The timer's timeout value is broadcast on the BCCH and covers a range of 6 minutes to about 25 hours. The timer is reset every time signalling takes place between the MS and BS.

Periodic location update is a good way of maintaining an accurate log of mobiles that are active. A benefit of the periodic location update feature is that it limits the time that an MS is not registered with the VLR, following a loss in the VLR transient database. As the LAI would remain the same after VLR recovery, a location update due to LAI would not take place: hence the benefit of the periodic feature.

8.5.6   IMSI attach/detach

When an MS is switched off, or the SIM removed, the mobile sends an IMSI detach message to the network. If the MS returns to the active state it performs an IMSI attach procedure and registers. However, if the location area identified is different from that stored in the SIM, a standard location update procedure is invoked. The IMSI attach/detach procedure is optional to the network operator. The feature, if required, is signalled to the MS over the BCCH.

8.6   Handover

8.6.1   Radio conditions and handover

Handover can be defined as the directed change of the radio channel used by the mobile to communicate with the network in order to maintain a good quality of service. A base station continuously monitors the performance of the link between the base station and mobile and instigates a handover to a better channel if required. For the uplink, this is achieved by using parameters measured by the base station; for the downlink the mobile takes the measurements and regularly reports them to the base station. The method of reporting mobile measurements for the purposes of handover is known as Mobile Assisted HandOver (MAHO). The important benefit in adopting MAHO, is that *actual* measurements of the radio environment around the mobile are used, rather than the base station making estimations based on the uplink only.

For the GSM reference model, a number of types of handover can be defined, e.g.:

- Intra cell handover - between traffic channels in the same cell
- Inter cell handover - between traffic channels in different cells
- Inter BSC handover - between traffic channels associated with BTSs connected to different BSCs
- Inter MSC - between traffic channels associated with different MSCs
- Inter PLMN - between traffic channels in different networks

The extent of involvement of the various network entities in the GSM network varies according to the type of handover and is covered in more detail below. The choice of handover is governed by the handover decision algorithm which operates on a number of rf parameters relating to the link.

8.6.1.1 Parameters for handover point evaluation

Measurements made by the mobile are sent to the base station over the SACCH, which can be associated with a TCH or SDCCH. The latter facilitates the handover process while the mobile is in the signalling state. A complete SACCH block of data is received by the base station every 480 ms, approximately. Each measurement is averaged over the SACCH block period. In addition to reports relevant to the downlink from the serving cell, the mobile also reports the received level of the six strongest surrounding, or neighbour cells, as well as their Base Station Identity Codes (BSIC), which can be decoded from the SCH on the BCCH of the neighbouring cells. The list of reported parameters is as follows:

  . Received signal strength downlink (RXLEV_DL)
  . Received signal quality downlink (RXQUAL_DL)
  . Downlink neighbour cell RXLEV (RXLEV(n)_DL)
  . BSIC of neighbour cell
  . Channel no?

While receiving the uplink measurements the base station would take its own measurements of the uplink. Main uplink parameters for a reliable handover evaluation are:

  . Received signal strength uplink (RXLEV_UL)
  . Received signal quality uplink (RXQUAL_UL)
  . MS-BTS distance (proportional to TIMING ADVANCE)
  . Power budget
  . Interference level in unallocated time slots

At the base station, the handover decision process involves an algorithm that uses the above information and runs in real time. Every mobile-to-base dedicated link requires a unique handover process to be run, thereby making the handover process quite demanding on base station resources.

In GSM the handover algorithm is left to the implementor to design, but a good example is given in Annex A of GSM 05.08. This forms the basis of the description in this section. It is based on the link characteristics, which are measured over a time window, and fed to a decision process, which indicates the instant when handover is desired. The link (and system) parameters considered to be service affecting are:

Signal strength: Poor reception of signals at the base station or mobile and the presence of a better link to another base station, or the same base station, but at a different frequency can result in a handover. As will be described, the base station can increase its transmit power, or order the mobile to increase its transmit power, to acceptable levels. If signal levels persist in being below threshold, in spite of such power control, an unconditional handover should take place.

Signal quality: Sometimes interference experienced in a channel can cause sufficient signal degradation that forward error correction schemes are unable to yield acceptable quality levels. Here a handover to a channel that is perceived to be better is recommended, even if the received level in the original is acceptable. However, if the received level is at the highest possible level (after power control) the handover requirement becomes unconditional. The effects of fading and noise on the rf parameters are averaged over an operator defined time window.

Mobile to base station distance: Depending on a network's radio planning, cell sizes do vary. The distance limit would have been entered in the base station's data base, and constantly checked against the actual distance, which is related to the timing advance. Exceeding the threshold warrants a handover to be performed unconditionally.

Power budget: Minimising transmit power is important in improving the overall performance of a network, because this results in the statistical reduction of interference levels in a PLMN. Hence it is helpful to select a channel with minimum path loss among the list of cells available. The parameter used in relating to path loss is known as power budget. A comparison of the received levels from the neighbouring cells' BCCH transmission, with the received level from the serving cell extrapolated to full power, is made.   In this way a measure of the effective transmission path loss is obtained, and a handover on the basis of power budget can take place, even if the received level and quality are acceptable.

Interference in unallocated timeslots: In order to select the best new channel among vacant channels on a new carrier, or even on the same carrier, the base station takes measurements of noise levels in free timeslots. This information is used in selecting the destination channel, in the case of handovers between carriers in the same cell, or BTS. It can also be used in the channel assignment process that precedes any structured procedure such as call set up.

Traffic reasons: A cell can alleviate congestion by handing over mobiles that are using a comparatively large timing advance to a neighbour. Such mobiles would be at the edge of the cell and therefore within the coverage of neighbouring cells. The instigation of handover due to traffic reasons would be controlled at the MSC, or BSC level. Another case in point can be found in microcellular networks that have an overlay macrocellular network, which was discussed in some detail above in Section 8.3.

8.6.1.2 Handover candidate selection

Apart from deciding on the instant when handover should take place, the handover algorithm should also create a candidate list of neighbour cells, which are suitable to take the call. This is done by estimating the path loss to each of the neighbour cells and ordering them in a preferred list that is made available to the MSC. The MSC selects the target cell from this list. The chosen cell may not be the one that offers the best rf path, however, because the MSC will also consider traffic loading at the cells in the selection process.

The candidate list is based on the evaluation of the following formulae:

RXLEV_NCELL(n) > RXLEV_MIN(n) + Max(0,Pa)
where Pa = (MS_TXPWR_MAX(n)-P)

This may be interpreted as meaning that the neighbour cell 'n' is suitable for handover if the received level is greater than the minimum defined level, provided that the MS has sufficient power capability. If the maximum power capability (P) of the MS is less than the maximum allowed transmit power, then this is corrected for in the value of Pa.

1(Min(MS_TXPWR_MAX,P)-RXLEV_DL-PWR_C_D)
-(Min(MS_TXPWR_MAX(n),P) -RXLEV_NCELL(n))-HO_MARGIN(n) > 0

Likewise this means that the maximum allowed transmit power for cell n is greater than the current cell's maximum allowed transmit power, by an amount which is greater than the difference between the downlink RXLEV from the cell n, and the RXLEV from the serving cell. In other words, there is no point in handing over to cell n if it needs a higher transmit power and if the rf channel attenuation to that cell is excessive. The maximum transmit power value allowed by the cell would be replaced by the maximum capability of the MS, if the latter is smaller. The comparison is skewed by the use of the hysteresis factor (HO_MARGIN) and the effects of power control (PWR_C_D). The parameters used above are defined again here:

RXLEV_NCELL(n) is the received level from cell n.
RXLEV_MIN(n) is the lower threshold of signal level associated with cell n below which handover to cell n is not allowed.
P is the maximum power a mobile can use (which varies with the classmark).
MS_TXPWR_MAX is the maximum power the mobile is allowed to use in the cell. This is broadcast by the cell at all times.
RXLEV_DL is the received signal strength in the downlink relative to the serving cell.
PWR_C_D is the difference between the maximum downlink RF power permitted in the cell and the actual downlink power due to the BSS power control.
MS_TXPWR_MAX(n) is the maximum power the mobile is allowed to use in the cell n. This is broadcast by the cell at all times.
HO_MARGIN(n) is the hysteresis factor - useful in preventing handover bounce.

The above parameters are based on *averaged* values of operator defined parameters, Hreqt and Hreqave, where:

- Hreqave is the number of measurements over which an average is computed
- Hreqt is the number of average results that are maintained

These procedures establish the preferred order of candidate target cells. These are candidates for target cells should a handover be required. The above relationships do not explicitly cause a handover to happen. In the event of a handover, the actual triggering of the handover process is determined by checking if the parameters, RXLEV, RXQUAL, MS-BTS distance, and power budget (PBGT), fall below defined thresholds. The BS continually checks a sample of values and recommends handover, with a cause relating to that parameter if a defined majority of samples relating to performance values fail the threshold test.

It is also worth noting at this point that handover is not only used in roaming, but also for achieving the best rf link within a cell, or for levelling out traffic loads if particular cells are overloaded.

Besides cell to cell *(inter cell)* handover there is also *intra cell* handover. Intra-cell handover occurs when a mobile changes channel on the same carrier to a different timeslot, or changes carrier, because the original channel may have been suffering from interference.

While the instigation of handovers due to system loading aspects are network originated, the results of the handover algorithm are still used for the purpose of ordering the candidate target cells.

8.6.2 Handover and synchronisation

While the radio subsystem decides on the need for handover as a consequence of radio link performance, the call control part of the base station manages the signalling sequence that results in the transfer of the mobile to a new channel. At a high level, the handover signalling scheme can be viewed as a procedure in which the serving base station informs the MSC of the need for handover and presents it with a list of candidate cells. The MSC, after selecting a suitable target cell, alerts the new base station and commands the mobile to change channel, by making a discrete change of traffic channels, once the uplink transmission is synchronised to the target base station's timing requirements.

The synchronisation of the mobile's uplink is an important part of the handover scenario. As described above the timing advance used by the MS is constantly updated by the serving cell to ensure that the bursts arrive at the BTS in the correct time window. The problem with handing over to a new base station is that the mobile must have obtained the correct phase information for the new channel before it starts communicating with the target base station on a dedicated channel. GSM offers four ways in which this can be achieved, defined as follows:

- non synchronous handover
- synchronous handover
- pseudo-synchronous handover
- pre-synchronous handover

In *non synchronous handover* the mobile first accesses the new base station over the random access channnel. The guard period for the random access burst is very much larger than that associated with the normal burst and as such does not require timing advance. The target base station responds with physical information about itself and provides the mobile with the value of timing advance it should use.

In *synchronous handover* the mobile has knowledge of the target base station's timing. A conceptually straightforward method to achieve this is to synchronise all the base stations in a PLMN or coverage area such that they all have the same absolute phase. A mobile can then reliably compute the required timing advance for the target cell by observing the timing of the received BCCH frame structure. Recall that the uplink is always timed relative to the downlink. Time synchronisation would be at the superframe level, since the system has to account for both traffic and control channels, which use the 26 frame multiframe and 51 frame multiframe, respectively.

If the operator does not have a synchronous network it is still possible to improve upon the non-synchronous handover by using a GSM phase 2 feature called *pseudo-synchronous handover*. Here the mobile uses the following two measures of phase difference (between base stations) to compute the timing advance:

- Real Time Difference (RTD), as provided by the serving cell
- Observed Time Difference (OTD), as calculated by the mobile

The new timing advance (TA1) can be calculated from the old timing advance (TA0) with the following equation:

$$TA1 = OTD - RTD + TA0$$

A diagrammatic derivation of the above equation is given in Figure 8.14. The waveform in A shows a pulse launched from BS0 which is the source base station. It arrives t0 seconds later at the mobile as shown in waveform B. The pulses in waveforms C and D show the case for the target base station (BS1). From these four waveforms it is simple to verify the above equation.

It is worth noting that synchronous handover can be viewed as a special case of pseudo-synchronous handover by assuming RTD = 0.

In the case of *pre-synchronous handover* the mobile assumes that the timing advance value for the target cell is 0, which applies to the case when a mobile is geographically near the BTS site. This approach can be used in small microcells.

8.6.3 Handover signalling

The extent of signalling that takes place during a handover procedure depends very much on the type of handover that is required. At one end of the scale is the handover between timeslots on the same carrier, at the other end is the handover that involves MSCs in a different PLMN, for example.. In all cases, however, the detection scheme and signalling are designed to be robust and with minimal breaks in the traffic channel. GSM provides for a number of handover scenarios depending on the network entities involved, viz:

. Intra-cell intra-BSS handover
. Inter-cell inter-BSS handover
. Inter-cell intra-BSS handover, which includes

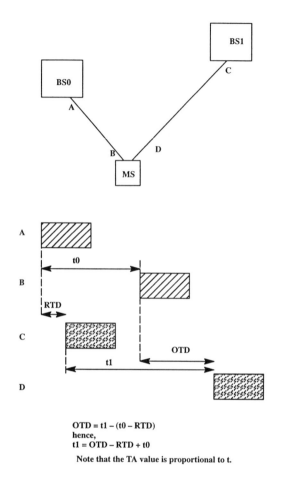

$$OTD = t1 - (t0 - RTD)$$
hence,
$$t1 = OTD - RTD + t0$$
Note that the TA value is proportional to t.

Figure 8.14   Timing advance computation with pseudo-synchronous handover

      inter-cell intra-BTS
      inter-cell inter-BTS

    Inter MSC handover, which includes
      inter MSC (MSC A to MSC B)
      inter MSC (MSC B to MSC A)
      inter MSC (MSC B to MSC C via MSC A)

Some of the main scenarios are now described

8.6.3.1 Intra-cell intra-BSS handover

In the simple case of intra-BSS handover (Figure 8.15), all the signalling takes place within the BSS, except at the end of the procedure, when the MSC is informed of the new channel parameters by means of the handover performed message.

The signalling does not have the handover related messages that are used in the other handover scenarios to be described. In fact, the intra-BSS handover signalling scheme borrows heavily from the call setup procedure in its use of the assignment command/response set. This avoids the unnecessary complications involving the MSC. Also timing calculations for the new channel are avoided because GSM 05.10 requires that all carrier structures in the same cell should be synchronised to the same time base counters and should be within 0.25 bit (12/13 μs) of each other.

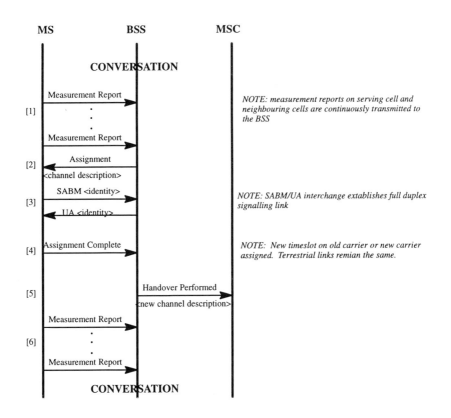

Figure 8.15   Message flow for intra-cell intra-BSS handover

Examining the diagrams, the main steps in the signalling sequence are:

[1] The measurement report is sent to the BSS every 480 ms enabling the handover algorithm to continuously update its status.

[2] If the channel in use deteriorates and the preferred alternative is another channel in the same cell the BSS sends an assignment message to the MS indicating the new channel description.

[3] A signalling link with the new channel is set up by bringing up the layer 2 link using the SABM/UA interchange.

[4] Over the signalling link (carried by the FACCH) an assignment complete message is sent to the BSS.

[5] The MSC is updated with the new radio channel description via the handover performed message.

[6] The call proceeds in the normal fashion using the new channel. Note that the measurement reports continue to be sent for the benefit of the operation of the handover algorithm.

8.6.3.2 Inter-cell inter-BSS handover

In the case of inter-BSS handover, the mechanism for identifying the need for the handover is the same as in the case of the intra-cell handover. However, the process of handing over to another cell results in a different signalling sequence, which is shown in Figure 8.16. The steps indicated in the diagram are described below:

[1] The measurement report is processed in the BSS in the normal fashion.

[2] The serving BSS recognises the need for a handover to another cell and issues a handover required message to the MSC while giving it a list of candidate target cells.

[3] The MSC selects the target BSS and sends the mobile details, such as the TMSI and path details (Circuit Identity Code), over the handover request message.

[4] An acknowledgment is received from the new BSS with the HandOver Reference Number (HORN), which is later used to verify that the correct mobile gains access to the new channel.

[5] The MS is commanded to handover with the handover command message.

[6] The MS uses the RACH to send a handover access message, demanding a response from the new BSS, containing the timing advance to be used in the new channel.

[7] The physical information message sends the timing advance information to the MS, which then stops sending the access bursts.

[8] A layer 2 signalling link is set up on the FACCH, and the MS uses that link to send a handover complete message to the network.

[9] The traffic link from the MSC to the old BSS is cleared, after the new BSS sends the handover complete message to the MSC.

[10] The old BSS confirms the release of the terrestrial resources with the clear complete message.

8.6.3.3 Inter-cell intra-BSS handover

In the case where the handover takes place between cells, which are controlled by the same BSS, the MSC is only brought into the picture at the end of the procedure as with the intra-cell case. The procedure is similar to the above, except that the handover command is autonomously generated in the old cell before the usual handover access/physical information messages are transmitted in the new cell. At the end of the sequence, an assignment complete message is sent by the MS to the new BSS, which subsequently sends a handover performed to the MSC.

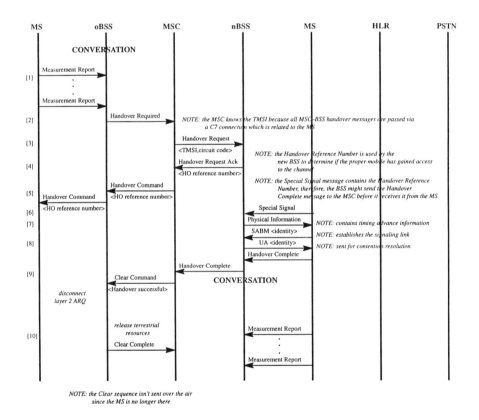

Figure 8.16   Message flow for inter-cell inter-BSS handover

The inter-cell intra-BSS handover can occur between BTSs controlled by the same BSC (inter-cell inter-BTS), or between cells controlled by the same BTS (inter-cell intra-BTS).

8.6.3.4 Inter-MSC handover

There are occasions when handover takes place between cells associated with different MSCs. This situation is recognised by the serving MSC, when the first suggested target cell in the candidate list received from the BSS belongs to another MSC. The sequence diagram in Figure 8.17 describes the message interchange in the case of a handover from MSC-A to MSC-B. The main steps to note are:

[1] After recognising that an inter-MSC handover is required, MSC-A requests a Handover Number (HON) from MSC-B by means of the prepare handover required message. This also provides the address of the target cell and the cipher key K_c. The HON is used to label the connection between MSC-A and MSC-B during handover.

[2] MSC-B requests the HON from its associated VLR.

[3] By sending a handover request message MSC-B forwards Kc to the target BSC and also requests a HORN and the provision of radio resource from the BSC. The HORN is used to relate the reserved radio resources in the target BSS with the MS to be handed over.

[4] The HON, HORN and details of the new radio channel are sent from MSC-B to MSC-A in the prepare handover response message.

[5] The Initial Address Message (IAM) and Address Complete Message (ACM) are used to set up a conference circuit in MSC-A, to ensure that the terrestrial network does not cause an interruption to the traffic path. The HON is used as routeing information to set up the connection to MSC-B.

[6] The MS is commanded to handover via the handover command message, which contains the HORN. This message also indicates the ciphering algorithm for the new cell.

[7] The first contact the MS makes with the new BSS is with the handover access message, which is sent over the RACH. It contains the HORN, which is used by the new BSS to link to the already assigned terrestrial channel in MSC-B. The successful reception of the handover access message results in a handover detect message being sent to MSC-B from BSS -B. This is transferred to MSC-A using the process access signalling request.

[8] The successful connection of the MS to BSS-B is signalled to MSC-A and BSS-A.

[9] Radio resources associated with the MS in BSS-A are released.

[10] MSC-A receives the answer signal from MSC-B after successful handover. It then initiates release of the conference circuit and its (MSC-A's) connection to the new BSS.

Other forms of inter-MSC handover include subsequent handover, when MSC-B hands back to MSC-A, and when MSC-B hands over to MSC-C, via MSC-A. These additional forms of handover involve some variations in the signalling, explained in GSM specification 03.09 [4].

8.6.4 Directed retry

GSM phase 2 has defined a procedure known as directed retry, which is essentially a handover in which the dedicated resource with a BTS in the signalling mode is handed over to another BTS, or cell, for the traffic mode. Figure 8.18 shows the case of an inter-BSS directed retry.

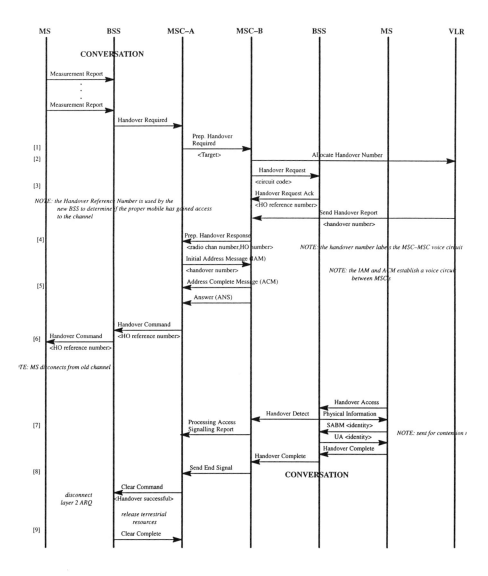

Figure 8.17   Message flow for inter-MSC handover

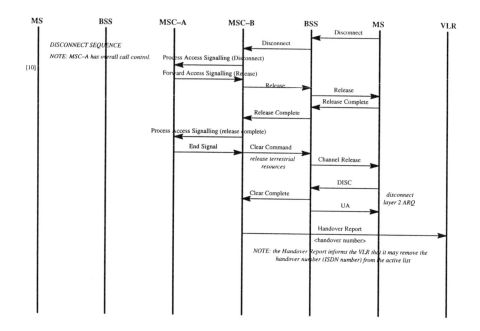

Figure 8.17   Message flow for inter-MSC handover (continued)

The main steps are:

[1] Towards the end of a call set up procedure an assignment request is sent to the BSS requesting a transition from the SDCCH to a TCH.

[2] If the BSS is experiencing congestion, or bad radio conditions, it can respond with an assignment failure with cause 'directed retry'.

[3] At this stage, the inter-cell inter-BSS standard handover signalling follows. Note that the cause values in the handover request and handover request messages will show 'directed retry'.

A simpler case is the intra-BSS directed retry. In this scenario the MSC is not involved in the handover control. A handover command requesting an SDCCH to TCH transition, where the TCH is controlled by another BTS, is generated by the BSC. After successful completion, an assignment complete message is sent to the MSC, indicating the new cell ID and with cause 'directed retry' in the message field.

8.7 Personal mobility aspects

The ability for a person to carry his/her cellular subscription capability, without a mobile phone, anywhere in a coverage area is referred to as personal mobility. In GSM, this is realised by a Subscriber Identity Module (SIM) card, which contains all the individual subscriber information. The SIM can be carried separately from the mobile equipment (ME).

The Authentication Key K_i, which is related to the IMSI, resides in the SIM instead of the mobile as a security feature. Also, a Personal Identity Number (PIN) facility is provided in order to protect against unauthorised use.

The subscriber can use the SIM with any mobile since the SIM-ME interface is standardised by GSM. When a SIM is removed, the mobile becomes unusable for normal services, but can still be used for emergency calls. Even though a number of different mobiles may be used by the same SIM, billing will always be attributed to the owner's account and the subscribed (paid up) facilities will be transferable to the MS in use, via the SIM.

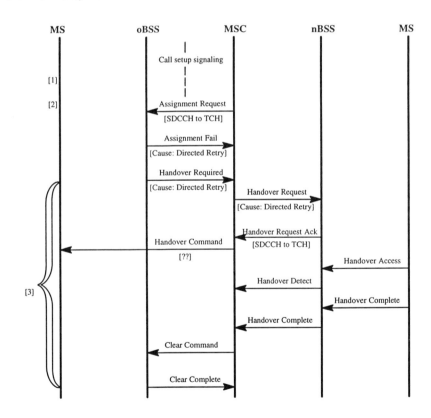

Figure 8.18   Inter-BSS directed retry procedure

8.8 Power control

It is not always necessary to transmit at full power to achieve reasonable link quality. Hence it is advantageous to alter the uplink and downlink power levels by a value sufficient to maintain acceptable quality. This feature is known as dynamic power control. The requirement to control power levels is designed to reduce the overall amount of interference from other cells, and to help conserve battery power in the MS.

Uplink power control refers to the mobile's transmission. Here the MS is ordered to change its transmit level by the BTS over a downlink signalling channel. The desired MS transmit level is decided in the BTS by considering the received signal level and received signal quality. The MS power levels can be reduced down to +5 dBm, in 2 dB steps, from a maximum level, determined by the class of MS. For example, a class 2 mobile has a maximum output power of 39 dBm (nominal), a class 5 mobile has a maximum output power of 29 dBm.

In downlink power control, the BTS transmit power level is decided locally based on the received level measurements, made by the MS, which are reported to the BTS within the measurement report. The BTS can have a range of 15 steps with a step size of 2 dB starting from its static rf power level. Note that the static power level is related to the power class of the BTS; for example, class 4 refers to a 40 W BTS, with the measurement taken at the input of the BSS transmit combiner. The specifications also allow for the static level to be adjusted by at least 6 steps of nominally 2 dB. This allows for power adjustment required as a result of rf cell planning. Implementation can be refined by fine control of the maximum power level in order to compensate for component tolerances.

Details of the MS and BTS static power levels and dynamic power level ranges can be found in GSM 05.05 Sec 4.1 [3].

8.9 Summary

A cellular system relies on mobility management to provide for the roaming capability, a key requirement. As cellular systems advance and new standards are created, mobility management has witnessed and will continue to witness significant improvements. GSM is the latest successful digital system and owes its success in part to a sophisticated mobility management philosophy, which embraces radio mobility, network mobility and personal mobility. These three aspects have been described in length in this chapter and their sophistication explained.

References

1   GSM 05.08 (ETS 300 578): 'European digital cellular telecommunications system (phase 2); Radio subsystem link control'

2   GSM 05.02 .(ETS 300 574): 'European digital cellular telecommunications system (phase 2); Multiplexing and multiple access on the radio path'

3   GSM 05.05 (ETS 300 578): 'European digital cellular telecommunications system (phase 2); Radio transmission and reception'

4   GSM 03.09 (ETS 300 527): 'European digital cellular telecommunications system (phase 2); Handover procedures'

Chapter 9

Strategies for mobility management

Amelia M Platt

Introduction

The principal characteristic of mobile networks, which distinguishes them from conventional fixed networks, is that the identity of the calling and called subscriber is not associated with a fixed geographical location. The subscribers establish a wireless connection with the nearest base station, or equivalent network access device, and can make and receive calls as they roam. Mobility management is concerned with how the network supports this function.

The cellular principle employed in mobile cellular networks, in which the geographical coverage of the network is divided into cells and each cell has its own BS[1], was proposed as a way of increasing the mobile network capacity [1]. In general, for a given coverage area, the smaller the cell size, the greater the network capacity, measured in terms of the total number of calls which can be established. Future cell sizes may be as small as the floor of a building [2] or even smaller, and thus the effect of subscriber roaming will be particularly significant since the cell boundaries will be crossed more often. Thus, while the cellular principle has increased potential network capacity it also increases the scale of the mobility management problem.

This chapter introduces the concept of mobility by differentiating between different types of mobility a network may provide. It then relates these to the mobility offered in current mobile cellular networks. A brief review of how mobility is managed in these networks is then given and is followed by a discussion of the costs incurred in managing mobility. However, the chapter is chiefly concerned with presenting the findings of studies into mobility management.

---

[1]   The principal components of the network are those described elsewhere in this book, but are also defined in Figure 9.1.

Finally, future trends in telecommunications and the implication of these on mobility management are discussed.

9.1 Types of mobility

In discussing mobility management it is necessary to distinguish between *terminal mobility* and *personal mobility*. Terminal mobility relates exclusively to mobile networks. In theory, terminal mobility allows a terminal to roam anywhere, provided it can communicate with a BS. Thus terminals are independent of any network access point and in general it is expected that the network access point will change regularly. Note that with terminal mobility, the association between the terminal and the subscriber is still retained; hence it can be said that first generation mobile networks provide only terminal mobility. The disadvantage of terminal mobility is that if a subscriber has more than one terminal then a caller must dial different numbers in order to reach the subscriber on the different terminals.

Personal mobility severs the link between subscribers and terminals and is applicable in both fixed and mobile networks. The network considers terminals and subscribers as distinct entities, each with unique identifiers and each requiring separate authentication. An association between a particular terminal and subscriber can be made known to the network, when required. Essentially subscribers register with the network and supply the identifier of the terminal they wish to use. This registration could be manual, or automatic, for example when a mobile terminal is switched on [3]. Hence, subscribers can access services at any terminal (fixed or mobile) of their choice.

Personal mobility also enables more than one subscriber to register at the same terminal. For instance, two subscribers may choose to receive calls on the same telephone. Conversely a subscriber can register to use more than one terminal at any time. Subscribers' identifiers will be accessible from any fixed/mobile terminal on a global basis, thus achieving the concept of 'anytime, anywhere', whatever the communication medium. This is the vision of third (next) generation personal communications networks.

In GSM, a subscriber personalises a terminal by inserting a personalised smartcard into the terminal. This allows calls to be delivered to an individual subscriber at any GSM terminal. Therefore, at any time, a GSM terminal can make/receive calls for, at most, one subscriber. Note that personalising a terminal is not the same as the personal mobility described above.

In summary, terminal mobility enables the terminal to roam around and across networks, while personal mobility enables a subscriber to roam between terminals and, if desired, be associated with more than one terminal and to share terminals with other subscribers. A detailed discussion of the concepts of terminal mobility and personal mobility, independent of any network architecture, is given in [4] and a possible implementation of personal mobility is presented in [5]. The main focus here, however, is the problem of keeping track of an entity (terminal or subscriber), a problem which exists with both terminal and personal mobility. Thus the information presented here is valid for both types of mobility.

9.2 Brief review of the cellular network architecture

Cellular networks have two distinct parts, a mobile part and a fixed part. The fixed part consists of BSs and MSCs. The MSCs are interconnected, typically using leased lines, and form the network backbone. MSCs are connected to BSs, with one MSC having responsibility for a number of BSs[2]. The mobile part consists of MTs. MTs and BSs have a radio interface through which they communicate. Interconnection of separate networks is usually achieved via public networks, for example, the PSTN.

To deliver an in-coming call to a subscriber, the network must know the subscriber's current location. This (and other) information is stored in the HLR and must be updated as the subscriber changes location. There is nominally one record in the HLR for each subscriber and, with the exception of the current location area, the information is fairly static. GSM and IS-41C (the European and North American standards respectively) define the HLR as a logical entity. In particular, the location of the HLR is not defined, and thus the distribution of the HLR will be network dependent.

VLRs are maintained by the network and are used to store location information about visitors who are currently active in the location area served by the VLR. The records are created when a visitor enters the location area and deleted when they move on. Note that the HLR maintains a pointer to the general area in which the subscriber is located (current location area), while the VLR maintains a pointer to the exact location (cell) of the subscriber. Similar to the HLR, the VLR is also defined as a logical entity, and thus the VLR location is network dependent. In practice, VLRs may operate as stand-alone units, serving multiple location areas [6,7] where a location area is defined as the granularity at which the network tracks subscribers. Alternatively they may serve one location area and are then generally co-located with MSCs [8]. While the physical location of the VLR does not compromise its functionality, studies indicate that the physical location of the VLR has a major impact on performance; this is discussed later.

Figure 9.1 shows a network architecture in which the VLR is co-located with the MSC and the HLR is distributed over the MSCs [7]. Other architectures are of course possible. Note that mobility management functions are considered natural Intelligent Network (IN) functions and possible ways in which the HLR and VLR databases could be included in the IN architecture are given in [3,9].

---

2   In GSM, base station controller parts are often shown separately from the MSC.

9.3   Cost of mobility

When a call is made to a mobile subscriber, the network must be able to locate the subscriber. In the literature many terms are used to describe this operation; we have adopted the term *find*, taken from [10]. Similarly, an *update* operation [10] is when a subscriber moves location and the change is recorded in the HLR. Thus the *find* and *update* operations are the two fundamental operations required to achieve mobility. There are two costs associated with both of them, namely communications and database processing costs. Note that the costs are incurred in the signalling network.

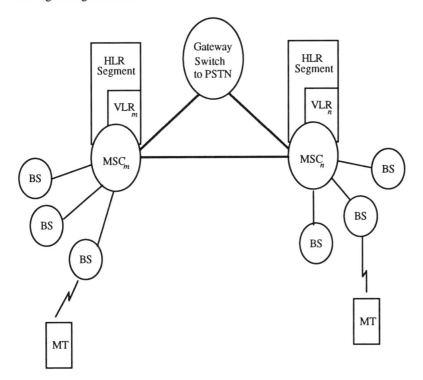

Figure 9.1   *Current cellular architecture*
[MSC=mobile switching centre, HLR=home location register,
VLR=visiting location register, BS = base station,
MT= mobile terminal]

9.3.1 Communication cost

The communication cost is directly related to the network resources required to transport the signalling messages and is a function of the number of hops in the route from source to destination. In packet switched networks the resources used at each hop are transmission bandwidth, processing power (to process the signalling message at the switch) and buffer space (to queue the message while awaiting transmission). Thus, in general, the communication cost is proportional to the end-to-end length of the connection.

9.3.2 Database processing cost

The resources used to process database transactions are disk channel capacity, processing power and memory. In general the cost of these is a function of the number of physical Input/Output (I/O) operations which must be performed in order to execute the transaction. Typically, read only operations require fewer physical I/O operations compared to write operations, because records have not changed and therefore do not need to be written back. The physical structure of the file and the file size also have a bearing on the number of I/O operations required. For example, a hashed file structure can be designed to give a 99% guarantee that a record can be read or written in one I/O operation (although a substantial disk space overhead is required to achieve this performance). In contrast the same operation using an index typically requires multiple I/O operations, with the exact number being dependent on the file size, type of index and various other parameters. A detailed discussion of physical file design and the various costs associated with file structures can be found in [11].

In addition to the basic cost of processing the database transactions, other database costs are incurred. For example, overheads are incurred in maintaining logging information required for database recovery. The exact costs are dependent on the DataBase Management System (DBMS) being used.

The delay associated with the *find* operation is also important, because this is the major contribution to the total delay experienced by the calling party at call set-up. The *find* delay is also a function of the number of hops in the route and number of I/O operations required. More complex calls will also tend to increase the call set-up time; hence efficient database design and distribution are critical.

9.3.3 Cost of mobility in GSM and IS-41

A detailed description of how the *find* and *update* operations are carried out in GSM and IS-41C is given in [5]. A simple analysis of the traffic generated by these operations for the VLR and HLR database accesses is also presented in this paper. Various assumptions are made, including the network size and call rates and, using the analysis, the query and update rates for the VLR and HLR are estimated. Note that the mobility *find* and *update* operations typically result in multiple database query and update operations. The detailed findings of this research are not reproduced here, but it is interesting to note that for both IS-41 and GSM, the total database traffic rates are the same, although the distribution across the HLR and VLR databases is different. In particular, there are more HLR queries per location area for IS-41 compared to GSM. This implies that as the network size increases, the location and distribution of the HLR is more critical for IS-41, than for GSM, if network congestion is to be avoided.

A framework for assessing signalling loads generated by mobile subscribers is also presented in [6] and subsequently used to compare the estimated signalling loads for IS-41 and GSM, given certain assumptions. Some of the findings were also reported in [12]. The study quantifies the traffic in terms of bytes per mobility operation, compared to database update and query rates per second considered in [5]. The main conclusions of these studies are that the total signalling load can be reduced by 50% for GSM, if the VLR is co-located with the MSC. Also, large switches capable of controlling multiple location areas reduce the signalling load, for both IS-41 and GSM.

Although the above analyses are useful in quantifying the signalling load associated with mobility, they do not give the total costs involved. In particular, the effect of the number of hops in the route on the communication costs and the database costs are not considered.

9.4   Mobility management research

It has been reported that the signalling load is expected to be much higher for cellular networks, compared to ISDN, and that this will increase further when a personal communication network is realised [12]. Also, most of the additional traffic is attributed to location updates. It is not surprising, therefore, that on-going research is attempting to find ways of reducing this cost. The aim of the section here is to consider some of the approaches which have been considered. Note that some proposals define a new network architecture, which is different from the architecture of current cellular networks. A reference which gave a survey and classification of strategies for locating subscribers was cited in [5].

9.4.1 Extreme mobility management strategies

Current practice is to update the HLR every time the subscriber changes location, so that the subscriber's location is always known. With this approach the cost of the location updates is maximised while the cost of finding a subscriber is minimised. Another approach is not to record location changes. With this scheme the cost of *update* is zero, but the cost of the *find* is maximised. These are termed the always-update and never-update strategies [10] and are extreme approaches in managing mobility. Note that there is a basic trade-off between the *update* and *find* operations and the parameters which affect this trade-off are call frequency and call mobility. The objective in mobility management is to minimise the overall cost and thus most schemes are a compromise between these two extreme approaches. Achieving this compromise efficiently, however, is not a simple task.

For instance, it could be argued that the always-update strategy is the most suitable for subscribers who are highly mobile so that their location is always known. However, if the subscriber does not receive many calls then the always-update strategy may not be the most efficient. In general the most efficient policy depends on the subscriber's call and mobility patterns and in the literature this is termed the call/mobility ratio [13]. The call/mobility ratio is defined as the rate of call arrivals (*finds*) to the rate of moves (*updates*). A number of proposals presented in the next section use the call/mobility ratio to decide how best to manage mobility on an individual subscriber basis.

9.4.2 Use of partitions

A tree-like network architecture, comprising a hierarchical structure of location servers with the BSs located at the leaf level of the tree, is proposed in [13] (location servers are essentially MSCs, which also incorporate the HLR and VLR functionality). Location servers are responsible for tracking subscribers currently resident in the subtree below. Subscribers are permanently registered with one location server, termed the home location server, and may also register as a visitor under another location server. Logical partitions are used to group location servers which are those most frequently visited by the subscriber. These are defined for each subscriber separately, by gathering data on the subscriber's mobility patterns. Location updates are required only when a subscriber moves into a new partition; thus optimal partitioning minimises the cost of updates. Note that higher level location servers are informed only of the partition in which the subscriber is located, not the exact location server. Finding a subscriber entails searching the tree and three candidate search strategies have been proposed. These search strategies are typical tree search strategies used in many computer science applications, and details of these and other search strategies can be found in [19].

A number of simulation studies have been carried out [13] to investigate the effectiveness of using partitions. The conclusion is that the use of partitions, obtained from subscribers' call and mobility profiles, reduces the overall cost of mobility management. Furthermore, particular search strategies are more efficient for particular call and mobility patterns.

9.4.3 Sophisticated search techniques

A similar tree-like architecture is proposed in [14]. At each level in the hierarchy, MSCs store location information relating to all the mobile stations in the subtree below, and thus the root MSC contains location information on all mobile terminals in the network. Three basic location management operations are defined: search, update and search-update. The aim was to develop protocols for efficient searches and updates of the hierarchy such that the number of messages resulting from location updates was reduced without increasing the number of messages required for searches. The scheme is essentially based on a sophisticated set of tree update and search algorithms. For example, location updates can be full updates, where all MSCs in the path from the source (old) and destination (new) BSs are updated. Alternatively, lazy updates can be executed, in which no MSCs are updated, but instead a forward pointer to the destination BS is kept at the source BS. A limited update is also defined, where the new location information is not propagated to all levels of the tree, but is restricted to the lower levels.

Searching for a mobile terminal is carried out by searching the tree, starting with the lowest level, until a MSC learns the whereabouts of the mobile terminal. Thus, in the extreme, a search as far as the root might have to take place. The scheme also includes a search-update operation whereby the location is updated after a successful search has taken place. Simulation results indicate that this reduces the search costs significantly. Clearly all variations of the basic search operations have different costs associated with them.

This proposal [14] is described as a static location management scheme, because an update/search-update strategy is chosen for each subscriber, and no attempt is made to change the strategy in response to changes in the subscribers' call and mobility patterns. This was subsequently enhanced to dynamically choose a location management scheme, based on past history of the subscribers call and mobility patterns [15]. Thus, over a period of time, a number of different update/search-update strategies may be in operation for the subscriber and the choice is based on the subscriber's predicted call and mobility patterns. This was termed dynamic location management and results indicate that the dynamic algorithm was more efficient.

9.4.4 Using forward pointers

A scheme, which is usable with the architecture more usually associated with cellular networks, in conjunction with forward pointers, is presented in [16]. When the mobile terminal moves location areas the HLR in not updated to reflect the new location. Instead, a pointer to the new location is placed in the old location area. Thus on searching to find the location of the subscriber a chain of pointers may be followed. At some stage the HLR will be updated, as otherwise the chain will continually grow in length, and also the chain of pointers will be removed from the VLRs. At what stage it is most appropriate to update the HLR depends on the cost of setting and traversing the chain of pointers, and this will depend on the call and mobility patterns of the subscribers. The conclusions are that forward pointers are beneficial for most call mobility profiles, except when the call/mobility ratio is low.

9.4.5 Replication

A scheme which is particularly useful for subscribers with a high call/mobility ratio
is presented in [17]. The proposal is to maintain a local store, or cache, of
subscriber location information (VLR addresses) at a switch, where a switch is
essentially a MSC. Location caching for a particular subscriber at a switch is
efficient only if a large number of calls for that subscriber originate from other
subscribers connected to the same switch . Hence a local call/mobility ratio, which
considers only the calls originating from that switch, is defined. The efficiency of
caching is a function of the probability that the cached pointer correctly points to
the subscriber's location and the probability increases as the subscriber's local
call/mobility ratio increases. Essentially the more calls which originate for a
subscriber from a switch, the greater the advantage in caching the subscriber's
location. To locate a subscriber, a cachefind operation which searches the location
cache is executed. A normal find is carried out if the cachefind is unsuccessful. The
cached pointer is updated when a full search must be performed. Essentially the
caching strategy distributes the functionality of the HLR to the switches and is
therefore beneficial if the HLR is likely to become overloaded.

9.4.6 Use of the wireless link to track mobiles

A novel scheme which uses the wireless link for tracking mobiles is proposed in
[10]. A subset of cells are chosen to be reporting cells. Location updates occur
only when mobile terminals cross into reporting cells. A vicinity is defined to be all
the cells surrounding a reporting cell up to other reporting cells, and therefore a
cell could be associated with more than one vicinity. To find a subscriber, the
vicinity of the last reporting cell entered is searched. Clearly, the cost of the find
operation increases with the size of the vicinity and the cost of the update increases
with the number of times a subscriber enters a reporting cell. Thus the problem is
one of selecting reporting cells such that the size of the largest vicinity and the total
cost of updates are minimised.

9.4.7 Distributed call processing

A distributed control architecture is proposed in [7] as an efficient solution to the
mobility management problem. In defining the architecture, the objectives were to
minimise the signalling load and call establishment times. The architecture
distributes both the processing and data required to manage mobility and deliver
services. User processes resident in the network carry out most of the processing
normally executed by the mobile terminals, and hence the signalling load on the air
interface is minimised. Also the call processing normally provided in the MSCs in
current mobile cellular networks is distributed over a number of servers. Home

location servers track mobile terminals currently operating in their home network, while visitor location servers track roaming mobile terminals.

The distributed control architecture has a number of significant benefits over current mobile cellular architectures. For example, call establishment time is reduced because service delivery is controlled by the home network. Therefore service control can be carried out in parallel with locating the subscriber. In contrast, in current architectures, the service control is provided by the terminating MSC. Therefore the subscriber must be located *before* the service control can be executed. In addition, only one database access is required to locate the subscriber. Call establishment time will become increasingly important as more complex calls, for example, multi-cast, multi-media, etc., become a reality.

Another advantage of controlling the call from the home network is that the subscriber receives a consistent service, only relying on the terminating network to provide transport services. Finally, in current cellular networks, terminating MSCs maintain the call state for active mobile terminals and therefore must remain in the connection for the call duration. This may result in inefficient routeing if a mobile terminal moves during the call. With longer call durations associated with multi-media calls, this will become an increasing disadvantage and may compromise the quality of service requirements of the call. For example the end-to-end delay negotiated with the network at call set-up may be exceeded. This is not a problem for the proposed distributed architecture, because the call state is maintained by a server and the call route can be modified, if necessary.

9.4.8 Findings

The various mobility management schemes summarised here are quite different in their approach to mobility management. However, a number of similarities are evident. A common objective is to minimise the overall cost of mobility support, and thus the trade-off in the cost of the *find* operation compared to the cost of the *update* operation is a recurring theme. Related to this is the subscriber's call and mobility pattern. A number of proposals, which have been discussed, are designed specifically to exploit this and mean mobility is managed individually for each subscriber. Thus subscriber call and mobility data is a valuable asset and a method of collecting and analysing this data will be needed.

The proposals which require the existence of a hierarchical architecture are restrictive. While telephone network architectures have traditionally been hierarchical, there is substantial evidence that the trend is towards flatter, mesh architectures. For instance, the network structure being deployed by SIRTI, the Italian domestic network operator, is non hierarchical [18].

Finally, the flexibility of the distributed call processing approach is an advantage and has several benefits. Combined with the exploitation of the subscribers' call and mobility patterns, distributed call processing could prove to be an efficient and robust solution to mobility management.

9.5   The implication of future trends on mobility management

It is now the case that in the US alone, tens of millions of people subscribe to mobile networks and this trend will undoubtedly be repeated globally. It has already been noted that the signalling load will likewise increase dramatically and that much of the extra traffic will be due to location updates [12]. Another reason for signalling to increase is that future mobile networks will support a much wider diversity of applications. For example, multi-cast, multi-media applications are anticipated [20], and more resources will be required to establish and control these types of calls. Reliability must therefore be a major objective for network providers. Mobile networks are particularly vulnerable to the loss of data stored in the HLR and VLRs. If these databases are unavailable then services cannot be delivered and thus the network will be considered to be effectively 'down' (even though the network infrastructure of links and switches may be functioning correctly). Figures reproduced in [21] indicate that high availability of these databases is a critical requirement and a maximum of only three minutes downtime per year can be tolerated. Database performance requirements are also quite stringent. For example, response times of less than 150 ms are required for 98% of all database queries.

A system in which critical data are distributed could satisfy some of these performance requirements more easily than if the data were centralised. Two of the key characteristics of distributed systems are scalability and reliability. In centralised systems, scalability of certain resources, for example processors, is limited. In distributed systems this limitation does not exist. With the expected increase in mobility, scalability will become an issue. In distributed systems reliability is achieved by replicating data and other resources and by providing software recovery procedures [22, 23]. The level of replication is tailored to satisfy the system requirements and in general will deliver enhanced performance and high availability, precisely the requirements reported in [21]. With centralised systems, reliability is typically achieved by deploying mated pairs [26].   Under normal operating conditions the mated pair shares the load;   however, each is engineered to carry the full load in the event of a failure.   This is a costly solution, requiring the complete system to be duplicated (hardware, software and data).

However, distributed systems also have a number of disadvantages. Clearly performance enhancements are only possible if the replicated data is placed judiciously. The critical effects of the VLR location have already been discussed. Distributing data is not an easy task anyway [24], and is even more complicated when the source and destination of the queries are constantly relocating. Also, replicated data is not easy to manage and requires much more control [25] compared to non-replicated data. All-in-all there is still much to be decided about how to efficiently manage mobility in mobile cellular networks.

References

1   MacDonald, V.H., 'The cellular concept', *Bell Systems Technical Journal*, vol 58, no 1, Jan 1979, pp 15-41

2   Steele, R., 'Speech codecs for personal communications',
    IEEE Communications Magazine, vol 31, no 11, Nov 1993, pp 76-83

3   Jabbari, B., 'Intelligent network concepts in mobile communications', *IEEE Communications Magazine*, vol 30, no 2, Feb 1992, pp 64-69

4   Zaid, M., 'Personal mobility in PCS', *IEEE Personal Communications*, vol 1, no 4, Fourth Quarter 1994, pp 12-16

5   Mohan, S., and Jain, R., 'Two user location strategies for personal communication services', *IEEE Personal Communications*, vol 1, no 1, First Quarter 1994, pp 42-50

6   Pollini, G.P., Meier-Hellstern, K.S., and Goodman, D.J., 'Signaling traffic volume generated by mobile and personal communications', *IEEE Communications Magazine*, vol 33, no 6, June 1995, pp 60-65

7   La Porta, T.F., Veeraraghavan, M., Treventi, P.A., and Ramjee, R., 'Distributed call processing for personal communications services', *IEEE Communications Magazine*, vol 33, no 6, June 1995, pp 66-75

8   Mouly, M., and Pautet, M-B., *The GSM system for mobile communications*, Palaiseau, 1992

9   Laitinen, M., Rantala, J., ' Integration of intelligent network services into future GSM networks', *IEEE Communications Magazine*, vol 33, no 6, June 1995, pp 76-86

10 Bar-Noy, A., Kessler, I., 'Tracking mobile users in wireless communications networks', *IEEE Transactions on Information Theory*, vol 39, iss 6, Nov 93, pp 1877-86

11   Folk, M.J., Zoellick B., *File structures*, Addison-Wesley, 1992, 2nd Edition

12 Meier-Hellstern, K S., and Alonso, E., Chapter 4, 'The use of SS7 and GSM to support high density personal communications', from *Wireless Communications, Future Directions*, Kluwer Academic Publishers, 1993, pp 55-68 (Based on the 'Third Workshop on Third Generation Wireless Information Networks' held in April 92 at Rutgers University)

13 Badrinath, B.R., Imielinski,   T., and Virmani, A., 'Locating strategies for personal communication networks', IEEE GLOBECOM 92 Workshop on Networking of Personal Communications Applications, December 1992

14 Krishna, P., Vaidya, N. H., and Pradhan, D. K., 'Location management in distributed mobile environments', *Proceedings of the Third International Conference on Parallel and Distributed Information Systems*, pp 81-88, Sept 94, Austin, Texas

15 Krishna, P., Vaidya, N.H., and Pradhan, D.K., *Static and dynamic location management in distributed mobile environments*, Technical Report 94-030, Dept of Computer Sc, Texas A&M University, June 1994

16 Krishna,, P., Vaidya,, N.H., and Pradhan, D.K., *Forward pointers for efficient location management*, Technical Report 94-061, Dept of Computer Sc, Texas A&M University, Sept 94

17 Jain, R., Lin, Y., Lo, C., and Mohan, S., 'A caching strategy to reduce network impacts of PCS', *IEEE Journal on Selected Areas in Communications*, vol. 12, no 8, Oct 1994, pp 1434-1444

18 Cancer, E., McCann, R., and Aboudaram, M., 'IN rollout in Europe', *IEEE Communications Magazine*, vol 31, no 3, Mar 1993, pp 38-47

19   Aho, A.V., Hopcroft, J.E., Ullman J.D., *Data structures and algorithms*, Addison-Wesley, 1983

20 Ychaudhuri, D., and Wilson, N., Chapter 17, 'Multi-media personal communications networks (PCN) : system design issues', from *Wireless Communications, Future Directions*, Kluwer Academic Publishers, 1993, pp 289-304 (Based on the 'Third Workshop on Third Generation Wireless Information Networks' held in April 92 at Rutgers University)

21 Wirth, P. E., 'Teletraffic implications of database architectures in mobile and personal communications', *IEEE Communications Magazine*, vol 33, no 6, June 1995, pp 54-59

22 Cristian, H, 'Understanding fault-tolerant distributed systems', *Communications of the ACM*, vol 34, no. 2, Feb 1991, pp 56-78

23 Coulouris, G., Dollimore, J., Kindberg, T., *Distributed systems concepts and design*, Addison-Wesley, 2nd Edition, 1994

24 Bell, D., Grimson, J., Chapter 4, 'Data handling - distribution and transformation in distributed database systems', from *Distributed Database Systems*, Addison-Wesley, 1992, pp 92-121

25 Paris, J., 'The management of replicated data', in *Hardware and Software Architectures for Fault Tolerance, Experiences and Perspectives*, Lecture Notes in Computer Science 774, Springer-Verlag, 1994, pp 305-311

26 Chow, T.S., Hornback, B.H., Gauldin, M.A., Ljung, D.A., 'CCITT Signalling System No 7: The backbone for intelligent network services', *IEEE Global Telecommunications conference*, Tokyo, Nov 87, pp 1561-1565

Chapter 10

Microcellular
networks

Professor Raymond Steele

Introduction

This chapter is concerned with microcells and the teletraffic that a microcellular network is able to support. A prerequisite is to discuss the numerous types of cells that are available to a network designer, and the role that microcells have, or may be expected to have, in high capacity mobile networks. We recall that a cell is an area surrounding a base station (BS) site where communications between mobile stations (MSs) and their BS are of an acceptable quality . The cell dimensions are different for an isolated cell compared to a cell in a cluster where the cochannel interference effectively decreases cell size. Cells are arranged in clusters, with clusters tessellated. The bandwidth allocated by the regulatory body (the Radio Communications Agency in the UK) is reused in every cluster in order to increase the number of radio channels that can be accommodated on the network. The bandwidth assigned to the cell sites within a cluster is in practice complex. For simplicity we will assume the division to be equal, unless otherwise stated.

10.1 Cell types

Table 10.1 lists the names of different types of cells and their largest dimension. First generation analogue cellular radio systems, e.g. TACS, AMPS, NMT, use large cells, macrocells, and recently, minicells. Currently, second generation cellular systems use the same types of cells, although in some dense urban areas microcells are either under consideration or have been deployed. Cordless telecommunication systems, exemplified by CT2 and DECT, do use microcells. Telepoint networks have street microcells, and indoor cells are referred to as indoor microcells, or indoor nanocells, or indoor picocells, depending on their dimensions. The Universal Mobile Telecommunication System (UMTS), planned for deployment at the beginning of the next century, will use all the types of cells listed above. As a consequence there will generally be overlapping tiers of cells.

Table 10.1   Types of cells and their largest dimension

Type of cell	Largest dimension
Picocell (usually in buildings)	a few metres
Nanocell (usually in buildings)	up to 10 m
In-building microcell (office(s), floor(s))	up to 100 m
Nodal cell (high capacity network node)	up to 300 m
Street microcell (pedestrian mobiles)	10 - 400 m
Street microcell (vehicular mobiles)	300 m to 2 km
Minicell (pedestrian mobiles)	500 m to 3 km
Macrocell (cities and suburbia)	1 to 5 km
Large cell (suburbia and rural)	5 to 35 km
Megacell (covering cities to counties)	20 to 100 km
Satellite cell (for global communications)	>300 km

10.1.1   Cell shape and size

A satellite beam on the surface of the earth determines the shape and size of satellite ground cells. Only massive terrain variations, e.g. mountains, will affect the cell shape formed by satellite spot beams, although there will be shadowing effects from large buildings and other man-made structures. Large terrestrial cells are dependent on the terrain of the earth, and because the BS antennas are mounted on the tops of tall buildings, or hill tops, the cell shape and size will depend on the transmitted power, terrain variations and the presence of large buildings. Cities are often modelled as 'urban clutter' when large cells or macrocells are designed.

As we make cells smaller we increasingly rely on local features to control the shape of the cell and the radiated power to control its size. The size and shape of street microcells are determined essentially by the streets and neighbouring buildings and only marginally depend on the terrain variations. Indoor microcells depend on the layout of the building, the materials used, the contents of the rooms, and the adjacent buildings. We emphasise that in microcells the differences in the heights of the BS and MS antennas are very small compared to those in larger cells, and that the radiated power levels are concomitantly smaller. The BS antennas in street microcells are located below the urban sky-line.

10.1.2 Types of microcells

All cells are three-dimensional as radio signals are propagated in three dimensions. However, mobiles may be essentially in one dimension, as in highway microcells. City-street microcells are two-dimensional, as the terminals are either located in vehicles, or carried by pedestrians that move along the streets or pavements, respectively. If the city is composed of sky-scrapers then a microcell may embrace a number of floors and is therefore three-dimensional. The size of an indoor microcell may be as small as an office, or the size of a floor, or a number of floors. Microcellular BSs may be located within buildings, or electromagnetic signals may be radiated into buildings from external microcellular BSs. We will consider here highway, street and indoor microcells.

10.2 Highway microcells

Conceived by Steele and Prabhu, much research has been done on highway microcells because of their relative simplicity, and because highways support heavy vehicular traffic and therefore are a profitable environment for microcellular deployment. A highway microcell is a segment of a highway. A microcell can be formed by mounting a BS antenna on a bridge spanning the highway, or on a pole located on the central reservation. If a directive antenna is used the resulting cigar-shaped radiation pattern directed along the highway is able to provide the required coverage. Notice that there will be coverage behind the BS antenna due to the finite front-to-back gain ratio of the antenna.

In highways passing through rural areas the path loss law is approximately inversely proportional to d^4, where d is the distance between the BS and the MS. If the road curves via a cutting then there is a diffraction loss before the above path law applies once more. Two-cell clusters can be formed with highway microcells of length from 1 to 2 km. For a vehicular mobile travelling at 70 mph the handovers between highway microcellular BSs may occur every 30 to 60 seconds so in a typical telephone call there will be a number of handovers. This highlights an important point, namely that microcells must be conceived with an infrastructure that can handle many handovers occurring rapidly, and without a severe handover penalty being incurred.

The fast fading in highway microcells is Rician in flat fading channels, with the Rician parameter (= power in the dominant path divided by the power in scattered paths) varying between near-Gaussian to near-Rayleigh as a function of distance, and in a manner that is difficult to predict.

10.2.1 Spectral efficiency of highway microcells

Spectral efficiency in cellular radio may be defined as

$$\eta = \frac{A_{CT}}{S_T W} \quad \text{Erlangs}/\text{Hz}/\text{m}^2 \tag{10.1}$$

where A_{CT} is the total traffic carried by the network, in Erlangs, S_T is the total area covered by the network, and W is the allocated bandwidth. By noting that:

$$A_{CT} = CA_c \tag{10.2}$$

where C is the number of microcells in the highway microcellular network and A_C is the traffic carried by each microcellular BS, and

$$S_T = CS \tag{10.3}$$

with S as the area of each microcell, and

$$W = MNB \tag{10.4}$$

where M is the number of cells per cluster, N is the number of channels per BS, and B is the bandwidth of each channel, then

$$\eta = \frac{\rho}{SMB} \tag{10.5}$$

where ρ is the utilisation of each BS channel:

$$\rho = A_c / N \tag{10.6}$$

Notice that the smaller S, i.e. the smaller the area of the microcell, the greater η. Both M and B are dependent on the multiple access method, modulation, forward error coding (FEC), diversity, equalisation, and so forth. We emphasise that making the microcell size small has a significant beneficial effect on η, and in addition it renders the propagation more benign, in that a dominant path often exists resulting in Rician fading.

10.2.2 Digital transmission in highway microcells

Consider the highway microcellular clusters shown in Figure 10.1, where each cluster has M microcells of length L, equal to $L_f + L_b$. Consider the wanted BS to be BS_W, communicating with the indicated MS. This MS also experiences interference from BS_I, but not from BS_F, as directive antennas, whose radiation pattern is shown in the box in Figure 10.1, are assumed. The microcells shown shaded in the drawing use the same radio channels. For an MS at the edge of the cell, the cochannel distance is $(M+1)L-L_b$, while the distance between the MS and BS_W is L_f, where L_b and L_f are defined in the diagram. This is a worst case condition for the signal-to-interference ratio (SIR), which may be expressed as

$$SIR = \left[(M+1)\frac{L}{L_f} - \frac{L_b}{L_f} \right]^p \tag{10.7}$$

where p is the path loss exponent, typically four.

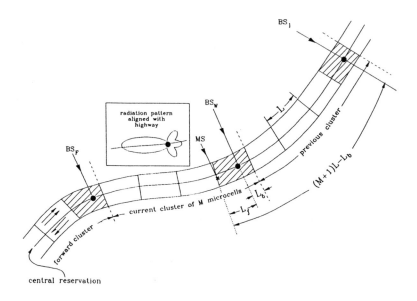

Figure 10.1   Highway microcellular clusters

There are n up-lanes and n down-lanes along the highway. For a particular average vehicular speed, each vehicle occupies a length along a lane of V giving $2nL/V$ vehicles in each highway microcell. Notice that V is a length of road that is longer the greater the vehicular speed. If $k(<1)$ of these vehicles are making a call, there are $2nkL/V$ active MSs. The bandwidth per cell is

$$\frac{W}{M} = 2Bn\frac{L}{V}k \tag{10.8}$$

giving

$$SIR = \left[\left(\frac{WV}{2nkB} + L - L_b \right) \frac{1}{L_f} \right]^4 \tag{10.9}$$

This equation relates bandwidths W and B with microcellular dimensions L, L_b and L_f, and with vehicular speed via V. Calculating specific situations can be very complex. Reference 4 calculates SIR and SNR and hence the overall probability of error for both Gaussian and Rayleigh fading channels. The calculations involve the use of Hermite polynomials. Graphs of bit error rate (BER) versus SNR for different SIR values are plotted in Reference 4.

10.2.3 Teletraffic issues

For most cellular radio networks there is no queuing. A call is either carried, or it is blocked and thereby lost. The grade-of-service (GOS) is the probability of blocking P_B. The offered traffic is

$$A = \lambda T_C = \lambda / \mu_c \qquad \text{Erlangs} \qquad (10.10)$$

λ is the average number of calls arriving during the mean call holding time of T_C. The reciprocal of T_C is the termination rate μ_c, also called departure rate and service rate. Also T_C has a negative exponential distribution. If the number of users is more than ten times the number of channels N at a BS, the Erlang-B formula is used, where the probability of a call being blocked is

$$P_B = \frac{A^N / N!}{\sum_{k=0}^{N} A^k / k!} \qquad (10.11)$$

Erlang-B tables show the relationship between P_B, A and N. It is important to note that for a given P_B (typically 2% is used), the offered traffic A is non-linearly related to N, giving a disproportionate increase in A as N is increased. The carried traffic is

$$A_c = (1 - P_B) A . \qquad (10.12)$$

When the number of users in a microcell becomes similar to the number of available channels, the Engset formula applies, where the probability that all the channels are busy is

$$P_N = \frac{\binom{K}{N} a^N}{\sum_{k=0}^{N} \binom{K}{N} a^k} \qquad (10.13)$$

and K is the number of mobiles in the cell and a is the offered traffic per user,

$$a = \frac{A}{K - A(1-P_B)} \tag{10.14}$$

Equation (10.13) is also called time congestion, i.e. the fraction of time when all the channels are busy. The probability that a call is attempted when there are no channels available is called call congestion, and this probability is given by

$$P_B = \frac{\binom{M-1}{N}a^N}{\sum_{k=0}^{N} \binom{M-1}{k}a^k} \tag{10.15}$$

Tables for the Engset formulae are readily available .

So far we have written equations that relate to large cells where the probability of handovers between BSs is low. When this is not so, as in highway microcells, the total call rate is

$$\lambda_T = \lambda_N + \lambda_H \tag{10.16}$$

where λ_N and λ_H are the new and handover rates, respectively. The channel holding time in a microcell may be a fraction of the call duration. A new call, i.e. a call starting in a microcell, has an average holding time of $\overline{T}_{Hn}$ while a MS entering a microcell, when already engaged in a call, has an average holding time of $\overline{T}_{Hh}$. Consequently the average holding time is

$$\overline{T}_H = \gamma_n \overline{T}_{Hn} + \gamma_h \overline{T}_{Hh} \tag{10.17}$$

where

$$\gamma_n = (1-\gamma_h) = \frac{carried\ new\ call\ traffic}{total\ carried\ traffic} \tag{10.18}$$

The carried traffic in microcells, where handovers frequently occur, depends on the blocking probability of new calls P_{bn} and on the probability of handover failure P_{fhm}, as well as $\overline{T}_H$. For non-priority schemes, where no priority is given to handover requests, the calculation is relatively simple, but for priority schemes it becomes complex and the reader is referred to References 7, 8 and 9.

10.3 City street microcells

In city streets the buildings are virtually continuous along the streets forming street canyons in which vehicular and pedestrian mobiles travel. The BS antennas are well below the sky-line and there is negligible diffraction over the roofs. There is considerable diffraction and reflections around buildings and there may be significant attenuation through narrow buildings into neighbouring streets. Diffraction losses experienced by mobiles travelling around corners is typically 20 dB.

As an MS travels along a street it receives signals arriving from different paths. Modelling the path loss has been done using two paths, one a direct line-of-sight (LOS), the other a ground reflected ray. Four path models include two extra paths from the buildings lining the street. The actual situation is more complex, but we can make some generalisations.

(1) As the MS moves away from the street microcellular BS the mean received signal level is relatively constant, and after some 10 to 20 m starts to decrease, often following a parabolic shape (signal level in dBs, versus log of distance in metres). It then may decrease more rapidly (4 to 6th path loss exponent). A break distance, d_b, has been observed, where before d_b the path loss exponent varies from 1.7 to 2.4, and after d_b from 4.5 to 9.2, where d_b varies from approximately 200 to 300 m for a propagation frequency of 900 MHz.

(2) The fast fading characteristics in a flat fading environment is Rician, and the variations of the Rician parameter with distance are extremely difficult to predict. Often a Gaussian-like channel occurs enabling the transmitter power to be decreased for a given microcell size and cochannel interference. However, this may mean that switching between BSs needs to be rapid as there is not the early warning associated with error bursts in Rayleigh fading channels.

(3) The observations stated in point (1) above are useful when analysing microcellular networks. However, measurements in many situations reveal a wide variety of path loss characteristics and estimating the path loss variations with distance is non-trivial. Consequently much effort has been invested in microcellular planning tools.

(4) Handover speeds are very much faster compared to conventional cells and many handovers may occur during a single call.

10.3.1 Outdoor microcellular prediction

The shape of microcells is primarily determined by the topology of the streets and the cross-sectional area of the buildings. If the city has rectilinear streets that are widely spaced, then siting a microcellular BS at a crossroad will yield a microcell in the shape of a cross. With the same transmitted power, but with the city-block size reduced, a more complex shaped microcell results. The electromagnetic energy now propagates down the four streets, and at each crossroad energy is propagated into them by a mixture of reflection and diffraction. The microcell now covers a number of city blocks.

For cities having complex road topologies and buildings, the shape of the microcell is to a first approximation the shape of the roads in the neighbourhood of the base site. However, the exact shape depends on the relative width and length of each road, and the cross-sectional area of each building in the vicinity of the BS. For example, the size of the microcell may be dramatically changed by relocating the position of an antenna from one side of a building to another . Open spaces, such as parks, do not offer the electromagnetic shielding that buildings do, although in some parts of the world the attenuation from foliage may be very high. The designer of street microcells does not look on buildings as urban clutter, but as a means to control and thereby shape the microcells. However, the control is often minimal, and if the size and shape of a microcell are not what is required, then the base site must be relocated. The design of microcells therefore requires a microcellular prediction planning tool that can rapidly predict the coverage of the microcells, as frequent re-siting of the BSs is required to achieve the required coverage and signal-to-interference ratios for the system being deployed.

Microcellular planning tools require a map scaling of about 1250 to 1 as prediction must be done on a per metre basis. A convenient approach is to use two-dimensional digital maps. This scaling provides details of roads and cross-sectional areas of buildings. No terrain information is available but this is not a problem, except in the case of escarpments, because the street microcell dimensions are small (with the largest overall dimension being typically less then 400 m). The microcellular planning tools are much more accurate than the crude radio area prediction tools used for macrocells because of the accuracy of the building cross-sectional areas used in microcellular predictions. For 85% of the predictions the prediction error is $< \pm 10$ dB.

Predictions made by MIDAS [11] start by ray tracing and once the model determines rays that are trapped within a street, they are grouped together as a super-ray which is then processed as a single ray. The attenuation experienced by this ray is dependent on a set of coefficients that have been selected from many propagation experiments. These coefficients depend on the street width, the regularity of the buildings on both sides of the street, the curvature of the street, the signal strength and so on. Once a side road or open space occurs, the super-ray is decomposed into a number of rays, which are all diffracted and reflected around the corner, using well established rules. Once around the corner they are re-grouped into another super-ray whose power is determined by the paths of the individual rays. The procedure is then repeated. Signal propagation across open spaces is modelled from a point using a circular wavefront that spreads out with distance with a particular loss per metre.

Figure 10.2 shows the predictions by MIDAS of an arbitrary and imaginary cluster of microcells in Central London. The colours represent different bands of path loss, with the smallest path loss in yellow, being next to the BSs. Because of space limitations, we are unable to display many figures, and so we make the following observations. BS LON1 overlooks a square and provides coverage over it and along the large roads that lead into the square. The signal in the roads one block behind (North East direction) LON1 are not well covered. LON2 offers a higher signal level over the southern tip of the square. The coverage from LON2 is basically a large cross. All BSs in Figure 10.2 use omnidirectional antennas, but when LON4 used a directional antenna pointing along the large road going North-West an effective one-dimensional microcell was produced, unlike the coverage indicated in Figure 10.2, where the lowest path loss in any location is shown. When the BS LON3 is pulled back into the small road behind it, the coverage across the park is triangular with the apex at the BS. There is no cochannel interference at LON2 due to LON9 and LON5. When LON9 and LON5 are assigned the same channel sets, the acceptable SIRs (say, 10 to 15 dB) mean that the effective microcell size is about half way between these BSs and the site of LON3. The deployment of the microcell cluster in Figure 10.2 shows areas of high path loss, necessitating the use of an oversailing minicell or a macrocell, to service mobiles entering these areas. Having an oversailing macrocell is desirable to facilitate handovers when candidate microcells have no available channels for this purpose.

10.3.2 Teletraffic in city microcells

Consider microcells tessellated in clusters with each cluster having M cells. Let each microcellular cluster be associated with a macrocell, and let the macrocells be formed into clusters having M_o macrocells per cluster, with the macrocellular clusters tessellated. Thus if the average area of a microcell is S, the average area of a microcellular cluster is MS (not to be confused with mobile station!), which is also the average area of a macrocell. The total area of this network, composed of one tier of microcells with an oversailing tier of macrocells, is MSM_o. For this discussion we will assume that the macrocells are there to support handovers (HOs) that cannot be handled by those microcells which currently have no available channels.

The total carried traffic is

$$A_{cT} = CA_c + C_M A_{CM} \tag{10.19}$$

where C and A_C were defined in connection with Eq.(10.2), C_M is the total number of macrocells in the network and A_{CM} is the traffic carried by each macrocellular BS. The channel utilisation for the network is

$$\rho = \frac{NA_c + A_{CM}}{MN + N_o} \tag{10.20}$$

where N and N_o are the number of channels at each microcellular BS and macrocellular BS, respectively. By using macrocells to assist with handovers the utilisation of the microcellular channels can be significantly increased .

10.3.3   Spectral efficiency

Applying the definition of spectral efficiency given by Eq. (10.1), and on noting that the total bandwidth is

$$B_r = B(MN + M_o N_o) \qquad (10.21)$$

the area of the network is given by Eq. (10.3), the total carried traffic is given by Eq. (10.19), and as $C = MC_M$,

$$\eta = \frac{A_{CM} + A_c M}{B(MN + M_o N_o)SM} \qquad (10.22)$$

10.3.4   Teletraffic simulation of microcellular networks

The teletraffic simulator, TELSIM, imports coverage predictions from MIDAS. The TELSIM operator then has to decide on the type of system to be deployed, e.g. GSM, DCS, IS-95, and for the chosen system select a set of appropriate parameters, such as fixed or dynamic channel allocation, the value of the processing gain, whether to use an oversailing macrocell, adaptive power control, intra- and inter-cellular handovers, SIR and SNR thresholds, handover hysteresis and timers, reserved handover channels, voice activity detection, multiple slot calls, adaptive modulation, and so on. Having set up the type of network, the coverage display imported from MIDAS, such as shown in Figure 10.2, is divided into a grid and in each square of the grid the relative number of users are assigned. Each mobile user is independent, offering the same average traffic that the operator types into the simulation menu.   Mobile users represented by small crosses move with uniformly distributed direction and Rayleigh distributed speed along the roads and in open spaces on the display. Figure 10.3 shows a snapshot (one frame) of a running simulation where the mobiles are connected by a line to their BSs. At handover the straight line connected to a MS switches from one BS to another. With the simulation in progress the mobiles are seen travelling along the streets and in open spaces. Notice in Figure 10.3 that there is an oversailing macrocell located at the top left hand corner of the display.

The simulation provides the number of calls logged; the mean and standard deviation of all calls, of completed calls, and of dropped calls; the new call blocking probability; the service probability; carried traffic per user; lost packet

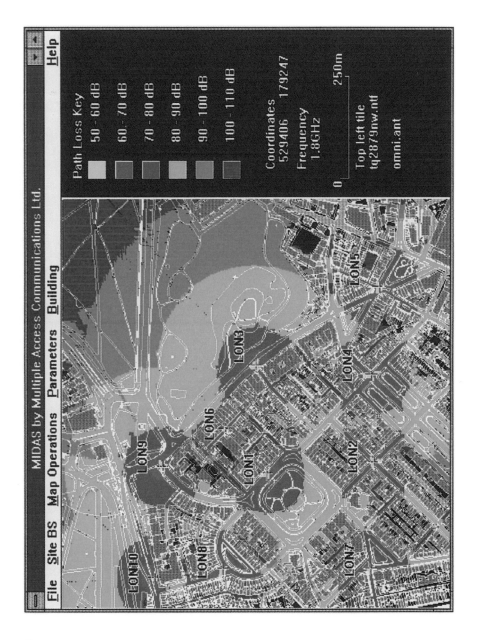

Figure 10.2   *MIDAS predictions of an arbitrary cluster of microcells in central London*
© Crown copyright

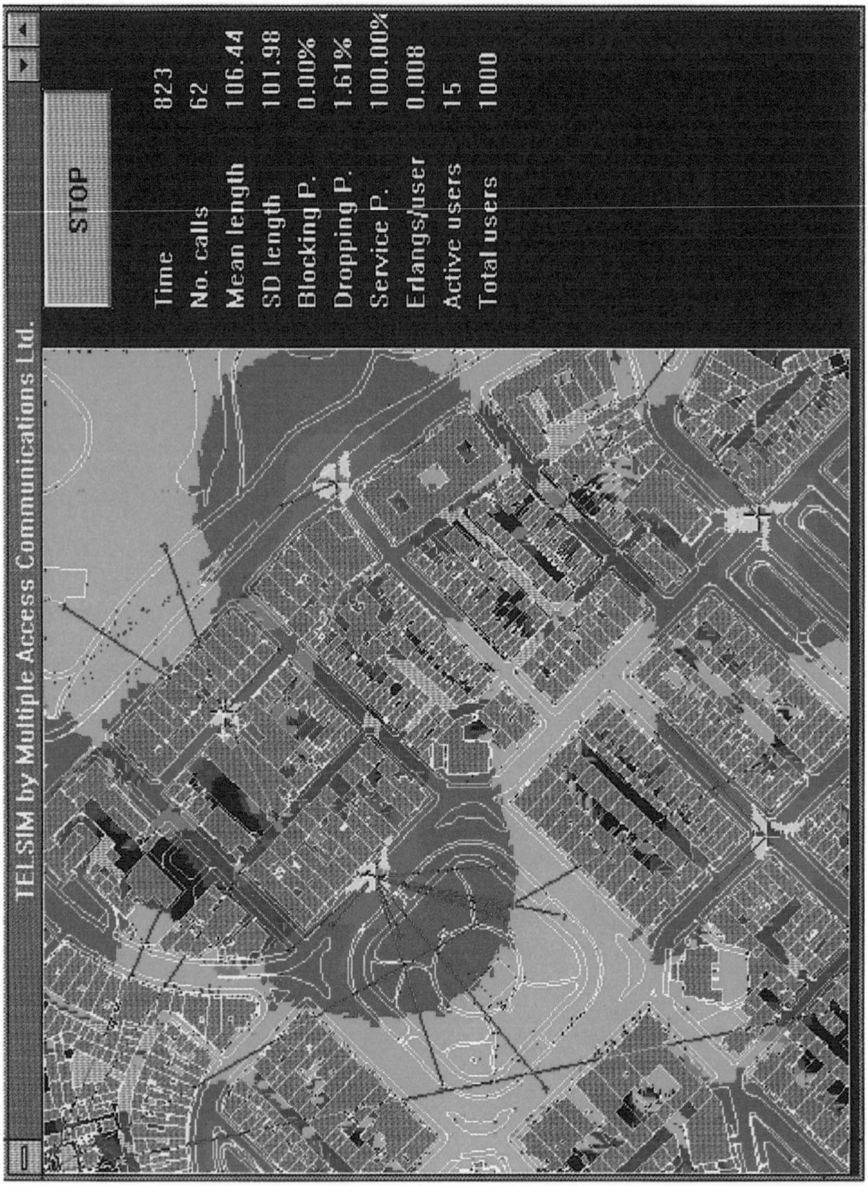

The display shows the following values:

Time	823
No. calls	62
Mean length	106.44
SD length	101.98
Blocking P.	0.00%
Dropping P.	1.61%
Service P.	100.00%
Erlangs/user	0.008
Active users	15
Total users	1000

TELSIM by Multiple Access Communications Ltd.

STOP

Figure 10.3   TELSIM simulation display using the cluster of microcells shown in Figure 10.2
© Crown copyright

rate; number of intercellular handovers per call; number of call attempts; percentage of new calls; channel utilisation; and so forth.

Armed with the microcellular planning tool and its companion simulator, we can make some surprising observations. As an example, let us consider GSM and DCS systems. They do not operate in simple cellular clusters when large cells are used, but to a first approximation, they have about 12 sectors in a reuse pattern. This may be viewed as a four cell cluster with three sectors per cell. When microcells are used the situation is significantly different. We have observed in three different cities using omnidirectional sites that a reuse factor of six for GSM900 and eight for DCS1800 is required.

In Reference 12 we simulated GSM and DCS in different microcellular clusters. The offered traffic per user was 12.5 milliErlang (mE), the mean call duration was 120 s, and the mean mobile speed 1 m/s. Each BS had one carrier, which therefore was a beacon carrier [1]. The peak transmission power of BSs and MSs was set to 10 mW. The noise floor of the MS was -100 dBm. To initiate a call, both the SNR and SIR had to exceed 12 dB, at both the MS and BS. When either the SNR or SIR <9 dB the dropping counter was implemented. If the counter was not reset because the SNR and SIR values did not improve over a 5 s period, the call was dropped. Every second the MS checked to see if it was being served by the BS offering the lowest path loss. If another BS had a path loss at least 3 dB lower, a handover request was made and a timer initiated. When the SIR <12 dB, the handover time was incremented and the MS changed its slot. The oversailing macrocellular sector had three carriers or 22 traffic channels.

The simulation was done for a city having rectilinear streets and for the centre of Southampton [12]. It was observed that the spectral efficiency was significantly higher in a rectilinear city than in Southampton for a two percent blocking probability. The number of BSs required for DCS1800 compared with GSM900 was higher by a factor of two for rectilinear streets, but only about 1.5 times for Southampton. Microcells in the centre of the group of microcells provided coverage, while those on the perimeter were much more heavily loaded. The macrocell enabled these microcells to operate near maximum channel utilisation because of its assistance with handovers. The macrocell employed only one sector for microcell support, there being four macrocells per cluster with three sectors per macrocell. As each sector had three carriers, the macrocell cluster utilised 36 carriers corresponding to 7.2 MHz. The microcells consumed 1.2 and 1.6 MHz of bandwidth for GSM900 and DCS1800, respectively. The spectral efficiency in E/MHz/km^2 for our 1 km by 1 km area was 2.55 for the macrocell only, 38.6 for the microcells only, and 7.9 for the complete GSM900 arrangement of microcells with an oversailing macrocell sector. The corresponding figures for DCS1800 were 3.15, 29.2 and 15.3.

10.4 Indoor microcells

These microcells may embrace a complete building or be the size of an office. In the same way that buildings in cities control the shape of a microcell once the antenna site and radiated power levels have been decided, so does the building construction and the building contents fashion the indoor microcell. As in street microcells, using simple equations to represent path loss can be extremely erroneous. Equations do have the virtue that research workers can use them to compare the performance of one system against another. In other words, the equations represent the path loss in a building environment that has a set of specific coefficients. An equation for path loss (*PL*) in dBs that is often quoted is

$$PL = L(v) + 20 log_{10} d + n_f a_f + n_w a_w \qquad (10.23)$$

where $L(v)$ is the mean of a log-normal distribution with variance v, and is a clutter loss that accounts for people, equipment, furniture, etc. [13]. The attenuation of the floors and walls in dBs is a_f and a_w, respectively, while the number of floors and walls along a straight line between the transmitter and receiver is n_f and n_w, respectively. Another representation is [14]

$$PL \begin{cases} 32 + 20 log_{10} d & ; \quad d < 25 m \\ 4 + 40 log_{10} d & ; \quad d \geq 25 m \end{cases} \qquad (10.24)$$

where the break distance, here 25 m, is environment dependent. Others quote that in the horizontal plane the propagation exponent is 3.5, and that wall losses could vary by large amounts (3 to 14 dB) at 1650 MHz. The range variations for a given path loss of 100 dB between floors could differ by factors of up to four. Line-of-sight propagation along a corridor may have an exponent of one (two for free-space) at 1.7 GHz, and one to three at 860 MHz . Scanning the literature reveals great discrepancies between measurements and this is due to the wide variety of buildings and their construction.

Fast fading in buildings is also very complex. An equation for the impulse response is

$$h(t) = \sum_{l=0}^{\infty} \left[\sum_{k=0}^{\infty} \beta_{k,l} e^{j\phi k,l} \delta(t - T_l - \tau_{k,l}) \right] \qquad (10.25)$$

where $\beta_{k,l}$ is the magnitude of the kth ray in the lth cluster of rays arriving at the receiver, and is Rayleigh distributed; $\phi_{k,l}$ is the phase angle of ray k,l and has a uniform PDF over $(0,2\pi)$; T_l , $l = 0, 1, 2,...$, is the arrival time of the lth cluster; $\tau_{k,l}$, $k = 0, 1, 2, ...$, is the arrival time of the kth ray measured from the beginning

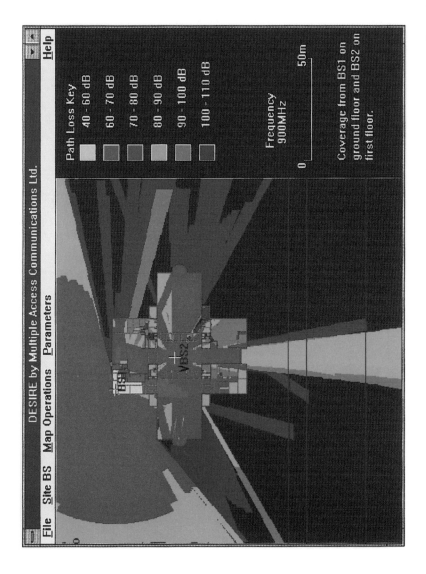

Figure 10.4 Coverage plot by DESIRE of the ground floor of a building

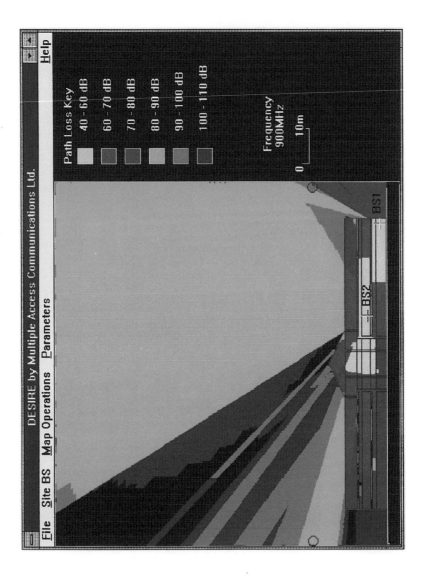

Figure 10.5   Side-elevation coverage plot by DESIRE for the microcellular base station shown in Figure 10.4

of the lth cluster; and T_l and $\tau_{k,l}$ are independent of each other, and their inter-arrival times have exponential PDFs. This model may be useful for simulations.

Again we urge caution in using rudimentary equations. The impulse responses measured by sounding a building are very complex, and may change rapidly and in surprising ways. This is because of the complex structure of the building, the variety of materials used, the building contents, and what can be most significant, the close location of other buildings that act as reflectors. The digital European cordless communication (DECT) system operates at a TDMA transmission rate of 1152 kb/s and a carrier frequency in the range of 1880 to 1900 MHz. Equalisers are not included in its design. However, we may suppose that its performance will be marginal in some indoor microcells. Equalisers are advised for transmission rates in excess of 2 Mb/s.

Abandoning the use of simple equations, the indoor microcell designer is required to rely on propagation measurements, or propagation prediction techniques. Radio propagation in buildings is much more difficult to accurately predict than in street microcells. That is the bad news. The good news is that it is usually much easier to site the indoor microcellular base stations and to change their positions within a building than in a street. If the building has low loss walls and floors, then shaping a microcell is very much dependent on the radiated power level and the position of the antenna sites. When the building has metal floors and walls, then it is easy to install microcells the size of an office as the office appears as a metal box. However, if say three offices are to appear as one microcell, then each office will have to have its own antenna and the three antennas will be connected to one microcellular BS. An alternative approach to antennas is to use radiating coaxial cables, and to run them above the ceilings to form microcells of the required size and shape.

A pre-requisite to predicting radio propagation in buildings is accurate drawings of the building. Next, the materials used in its construction must be identified. This is difficult because they often cannot be visually identified and therefore a knowledge of the principles of building construction is valuable. If it is known that the building has brick walls (visually apparent), then in the UK, for example, the bricks on the south-west part of the building will have a higher loss than those on the north-east face, because the prevailing winds are south-westerly. However, a guess at the wall attenuation can have a substantial error because the attenuation is dependent also on the carbon content of the bricks, which is not visually apparent. A radio planner is well advised to adopt the behaviour of a surveyor, lifting floor and ceiling tiles, tapping walls, and so forth.

The contents of buildings also present a challenge. A microcellular design must be sub-optimum so that movement of metal filing cabinets, or the introduction of a row of metal cupboards, will not produce radio dead-spots. Consequently, indoor planners must design for the path loss to be increased by the occupiers of the building long after the installation of the equipment has been completed. This means that microcellular prediction planning tools can be used even if the standard deviation of the prediction errors is large, say 8 dB.

An alternative to prediction tools is to use measurements. For example, first place the transmitter at a corner of the building and move the receiver to obtain the microcellular boundary. Next, exchange the position of the transmitter and receiver. The microcellular coverage now extends to the corner of the building and a considerable distance in other directions. Often one microcellular BS is insufficient for complete coverage, or more are required for capacity. Every time another BS is introduced, or its location adjusted, more measurements are required. There always seems to be a part of the building not covered, or unwanted radiation is leaking out of the building, so that a passer-by might be able to register on the indoor BS and this may be undesirable. Prediction planning tools therefore have a role to play, in that they are quicker and give results with sufficient accuracy for an acceptable design.

The basic method of indoor prediction is ray tracing in some form or other. Contents of the building are usually ignored. A menu giving the attenuations of the different walls, partitions and floors is required. These cannot be easily obtained and are usually based on either some measurements or previous experience. Figure 10.4 shows the ground floor of a large three-storey commercial building, whose outer walls were composed of metallised glass having transmission losses of 30 dB. However, the front door was made of plain low loss glass. One or more base stations were to be located within the building, such that they would provide coverage within the building for path loss up to 110 dB, while restricting the radiation from the front-door to prevent mobiles outside the building from attempting a handover to the indoor base stations. The signal level outside the building, due to an external base station (not shown), was such that the path loss from the in-building BSs to the exterior of the front of the building (south side of the building) was required to be greater than 80 dB for the power levels to be used by the indoor stations. Figure 10.4 shows base station BS1 located at the north-west corner on the ground floor. A virtual BS, VBS2, was located at the centre of the plan in order to predict the effect on this floor of an actual base station BS2, positioned on the first floor, as shown in Figure 10.5. Each colour in Figure 10.4 and Figure 10.5 represents a band of path loss. The coverage outside the building on the north-west side was permissible, as it was a company car-park, and calls were allowed to be made via the indoor BSs. Figure 10.5 shows that the path loss specification was met using two BSs, BS1 and BS2. The sites of these BSs were found after much resiting, demonstrating the necessity of a planning tool, rather than numerous and time consuming measurements for different base station positions.

10.5 Interconnecting microcellular base stations

While DECT and CT2 systems have microcellular BSs, cellular network operators are currently obliged to install full-size BSs when deploying a microcell. However, small-size BSs for cellular networks exist in development laboratories. Looking ahead, it is not clear what form microcellular BSs will take; they may be truly miniature versions of current BSs utilising advances in microelectronics, or at the other extreme they may be radio outposts known as Distribution Points (DPs) [18].

With DPs, the microwave radio signal that would normally be transmitted to mobiles, amplitude modulates a laser. The resulting optical signal is sent over an optical fibre to a DP having a photo-detector that recovers the radio signal. Microwave power amplification ensues and the signal is transmitted to the mobiles. This technique is known as radio-over-fibre. The received radio signal at the DP from the mobiles is low-noise amplified and then modulates a laser for the return link over the fibre to the location of the real BS where the microwave signal is recovered. Thus a base station controller (BSC) is co-sited with many BSs and their signals are distributed to DPs via optical fibres.

Should the approach be adopted of transmitting from a BSC the digital baseband mobile signals to small microcellular BSs, then the outcome will not necessarily be more expensive than the use of DPs. Microcellular BSs transmit low powers, typically 10 mW or so. The low radiated power levels considerably simplify BS design, and lower BS cost. Indeed, we anticipate that the cost of deploying a microcellular network will largely reside in the cost of interconnecting BSs, and not in the cost of the BSs. Finally, we emphasise that connecting microcellular BSs by means of radio, as in DCS1800 networks, should be avoided if microcellular networks are to carry vast amounts of teletraffic.

10.6   Multiple access methods

The spectral efficiency of a microcellular network is inversely proportional to the number of microcells per cluster, M: see Eqs. (10.5) and (10. 22). Fixed Channel Allocation (FCA) used with TDMA, as exemplified by GSM, DCS1800 and IS-54, yields cluster sizes considerably greater than one. For example, GSM can theoretically operate with three omnidirectional macrocells per cluster, but in practice four cells per cluster with sectorisation are used. Code division multiple access (CDMA) can operate with $M<1$ due to sectorisation gains. Even if Dynamic Channel Allocation (DCA) is used with TDMA, where cluster size has little meaning, the capacit y remains less with CDMA.

In microcells the conclusion as to what is the best multiple access method is not resolved. Buildings have a significant effect on cochannel interference. Figure 10.2 shows that microcells are irregular if the city has an irregular pattern of streets. A consequence is that cochannel interference may be substantial in a small part of a microcell. This means that rapid handover is required. This phenomenon also causes considerable difficulty in frequency planning in TDMA/FCA networks.

However, TDMA with DCA will fare much better, as can be shown by simulation. CDMA was conceived to operate in high cochannel environments and is therefore very suitable to combat the interference that appears in small areas of a microcell. However, for CDMA to achieve the necessary diversity gains from its RAKE receiver there must be sufficient multipaths occurring in the microcell. Consequently CDMA needs to transmit at a high chip rate in microcellular environments. By doing so it will not only enhance its immunity to cochannel interference from neighbouring cells, but will also support a wider range of services.

10.7   Conclusions

This chapter is designed to introduce the reader to microcells. We have discussed highway and city street microcells, and briefly introduced the more complex indoor microcells. Equations of spectral efficiency show it to be inversely proportional to the area of the microcell. The smaller the microcell the better the propagation environment and the better the frequency reuse, and hence the greater the spectral efficiency. For microcellular networks to be economical we need microcellular BSs that have minuscule costs compared to current ones. Indeed, microcellular BSs need to be less expensive than current cordless BSs. The methods of interconnecting microcellular BSs will be by optical fibres, and so the realisation of microcellular networks depends on the availability of low cost fibre and, of course, a suitable regulatory environment. A key factor in realising microcells is rapid handovers and preferably soft ones.

A brief discussion has been given regarding multiple access methods for microcells. However, it is only part of the picture. The real future lies in multimode terminals supporting Personal Digital Assistants (PDAs). Space prevents us from addressing the complex subject of the Intelligent Multimode Terminals (IMT) and their associated network which is complementary to microcells. The IMT is complementary to microcells, and the reader might like to read References 19 and 20 as a starting point.

Acknowledgements

The author thanks Multiple Access Communications Ltd for their permission to publish plots obtained from their software products, MIDAS, TELSIM and DESIRE. He also thanks his colleagues, John Williams, Shirin Dehghan and Aidan Collard. The author is grateful to Cellnet for permission to include the in-building coverage plots obtained using DESIRE.

References

1   Steele, R. *'Mobile radio communications'*. Pentech Press, London 1992, ISBN 07273-1406-8

2   Steele, R. 'Speech codecs for personal communications', *IEEE Communications Magazine*, vol 31, no 11, Nov 1993, pp 76-83

3   Steele, R. and Prabhu, V.K. 'Mobile radio cellular structures for high user density and large data rates', *Proc of the IEE, Opt F*, no 5, August 1985, pp 396-404

4   Wong, K.H.H. and Steele, R. 'Digital communications in highway microcellular structures', *IEEE Globecom*, Houston, USA, vol 2 of 3, 1st-4th Dec 1986, pp 31.2.1-21.2.5

5   Chia, S.T.S., Steele, R., Green E. and Baran, A. 'Propagation and bit error ratio measurements for a microcellar system', *J Inst of Electronic and Radio Engineers*, vol 57, no 6 (Supplement), Nov/Dec 1987, pp S255-S266

6   Boucher, J.R. *'Voice teletraffic systems engineering'*, Artech House, 1988

7   Hong, D. and Rappaport, S.S. 'Traffic model and performance analysis for cellular mobile radio telephone systems with prioritized and non prioritized hand-off procedures', *IEEE Trans Veh Tech*, vol VT-3, August 1986, pp 77-92

8   El Dolil, S.A., Wong, W.C. and Steele, R. 'Teletraffic performance of highway microcells with overlay macrocell', *IEEE Selected Areas in Communications*, vol 7, no 1, January 1989, pp 71-78

9   Steele, R and Nofal, M. 'Teletraffic performance of microcellular personal communication networks', *IEE Proc-I*, vol 139, no 4, August 1992, pp 448-461

10   Green, E. 'Radio link design for microcellular systems', *British Telecom Technology J*, vol 8, no 1, January 1990, pp 85-96

11   Webb, W.T. 'Sizing up the microcell for mobile radio communications', *Electronics & Communications Engineering J*, June 1993, pp 144-140

12   Steele, R., Williams, J., Changler, D., Dehghan, S and Collard, A. 'Teletraffic performance of GSM900/DCS1800 in street microcells', *IEEE Communications Magazine*, vol 33, no 3, March 1995, pp 102-108

13   Keenan, J.M. and Motley, A.J. 'Radio coverage in buildings', *British Telecom Tech J*, vol 8, no 1, Jan 1990, pp 19-24

14   Button, J.E. 'Power control in small cell CT2/CAI systems', *IEE Colloquium Digest n*o 1990/165, pp 5/1-5/6

15   Owen, F.C. and Pudney, C.D. 'In-building propagation at 900 MHz and 1650 MHz for digital cordless telephones', *6th Int Conf on Antennas and Proparation*, ICAP/89, Pt 2: Propagation, Conf Pub No 301, pp 276-280

16   Bultitude, R.J.C., Melancon, P. and LeBel, J.'Data regarding indoor propagation', Wireless '90, Calgary, Canada, July 1995

17   Saleh, A.A.M. and Valenzuela, R.A. 'A statistical model for indoor multipath propagation', *IEEE JSAC*, Feb 1987, pp 128-137

18   Wake, D., Westbrook, L.D., Walker, N.G. and Smith, I.C. 'Microwave and millimetre-wave radio over fibre', *British Telecom Tech J*, vol 11, no 2, April 1993, pp 76-88

19   Steele, R. and Williams, J.E.B.'Third generation PCN and the intelligent multimode mobile portable', *Electronics and Communications Engineering J*, vol 5, no 3, June 1993, pp 147-156

20   Steele, R. 'The evolution of personal communications', *IEEE Personal Communications*, vol 1, no 2, 2nd Quarter 1994, pp 6-11

Chapter 11

Future systems:
cellular access technology

Walter H W Tuttlebee

Introduction

Cellular telephony has grown since its inception in the 1980s at a phenomenal rate, with market growth commonly exceeding analysts' predictions. The transition from analogue to the new digital systems has provided further impetus to growth, as has the trend towards smaller, cheaper, personal handsets (replacing carphones as the main product) as well as the increasing regulatory liberalisation of European telecommunications markets. In North America, Personal Communications Services (PCS), based on a range of technologies, are being facilitated, and are seen to represent a major new market over the next few years. It is within such a context that Europe, and the world, are seeking to establish a route towards future, so-called, Third Generation Mobile Systems, for introduction early in the 21st century.

The Global System for Mobile communications, GSM, had, by the mid-1990s, been adopted by many nations, far beyond Europe, as the digital cellular telephony standard of choice; however, its preliminary specification stages began some 10 to 15 years earlier. Given such gestation timescales, technical R&D work in Europe on successor third generation systems has already begun. These systems are variously referred to as UMTS (Universal Mobile Telecommunications System), FPLMTS (Future Public Land Mobile Telecommunications Systems) or IMT-2000 (International Mobile Telecommunications, - the '2000' referring to the implementation timescale and frequency band).

This chapter describes the system requirements and technical issues which may influence the air interface and system access choices for such third generation personal communications, reflecting the focus of recent R&D activities in Europe. In particular, the chapter concentrates upon CDMA and TDMA air interface technologies. Key requirements that will influence the evolution of future cellular architectures and system design are outlined, along with a brief historical and commercial context. The chapter then describes the key system concepts and technologies of CDMA cellular radio and recent advances in TDMA. Progress in European CDMA and TDMA systems research, based around the CODIT and ATDMA testbeds, supported by the European Community RACE programme, and others, is also described.

11.1 General comments

Earlier chapters in this book describe many of the basic technologies of cellular radio systems. For further detail of the evolution from the early analogue cellular systems, AMPS, TACS and NMT through to today's digital TDMA systems such as GSM and DAMPS the reader should see Reference 1. Figure 11.1 provides a summary timeline showing the evolution of such systems, to provide a context for what follows.

In parallel with digital cellular telephony development, digital cordless systems for operation in microcellular environments have also been developed [2]. Future cellular systems will encompass the functionality of both service concepts and will draw on both sets of technologies.

In this chapter we examine the role that CDMA and advanced TDMA technologies may play as radio access technologies in tomorrow's systems. CDMA technology was considered in the early days of GSM specification, but at that time was deemed immature; however, for next generation systems CDMA is a potentially important technology.

Following a late recognition of European GSM developments in the US, a TDMA evolution of AMPS was standardised in the late 1980s, known as Digital AMPS or DAMPS or IS-54. However, the early 1990s saw a commercial direct sequence DS-CDMA cellular system promoted in the USA, as a competitor to the DAMPS system. Following field trials of this system, it has since been standardised, as IS-95, finally demonstrating the feasibility and performance capabilities of such systems. Other CDMA systems are now also being developed for cellular or RLL (radio local loop) applications.

The early UMTS air interface standards debate in Europe reflected the US standards arguments, i.e. pitting CDMA against TDMA. In fact the debate has matured as work on both CDMA and advanced TDMA has yielded many techniques, which could be used with either CDMA or TDMA. UMTS and FPLMTS will not simply be 'more-of-the-same cellular radio'. Services supported by UMTS will be much more inclusive and flexible than today's systems - personal communications and much more; the UMTS air interface will thus be required to accommodate efficiently multiple services in multiple environments, placing new technical demands on the air interface definition. Further, the envisaged market growth for radio based telecommunications services over the next two decades is huge, resulting in profound commercial pressures; it should be appreciated that, just as for GSM, whilst technical issues discussed in this chapter will influence the standards evolution, other commercial and political factors are likely to have at least as much influence.

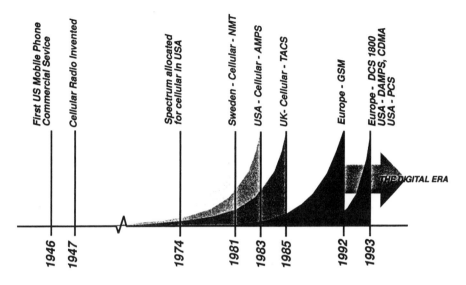

Figure 11.1   Evolution of mobile telephony standards

The chapter is therefore structured as follows.   We firstly discuss the requirements that are driving and constraining the development of future systems and architectures. We then review the principles and application of CDMA and advanced TDMA multiple access technologies to cellular radio.   Finally we describe four advanced hardware testbed systems which serve to illustrate implementation of some of the emerging techniques.   These three main sections overlap to some degree, due to the fact that many of the emerging techniques are closely related to fundamental principles and are driven by the new requirements.

Taken together, the chapter provides the reader with a broad understanding of future systems requirements and an insight into some of the relevant air interface techniques and trends. In this latter respect the chapter is incomplete in that it intentionally constrains its coverage to CDMA and TDMA, omitting discussion of other new technologies which may also impact upon third generation interface design, such as Orthogonal Frequency Division Multiplexing (OFDM), and Spatial Division Multiple Access (SDMA).   These technologies have emerged more recently and only limited work has been published to date on their application to third generation systems.

11.2   Requirements of future systems

All new systems are developed to meet certain desired design criteria, as was GSM for example.

UMTS will not be simply 'more-of-the-same cellular radio' - the regulatory, technological, service and operational environments in which UMTS will find itself imply demanding new design criteria, some of which are outlined below.

Generic service requirements: Second generation systems will not support the high rate services being launched on broadband fixed networks in the late 1990s, but UMTS will offer a subset of these.   It is not possible to predict at this time the specific services which third generation systems will offer, but we can identify some generic 'test case' services which must be supported - indeed this is necessary to allow such systems to be designed.   A set of such generic services are outlined in Table 11.1.

Table 11.1:   Example third generation generic service types

Service	Design constraint	Performance measure
Voice	Delay < 30 ms	MOS > 4.0 Decoded BER < 10^{-3} Frame error rate < 2%
Low delay data	Delay < 30 ms	BER < 10^{-6} Errored sec < 10 s/h
High delay data	Delay < 300 ms	BER < 10^{-6} Errored sec < 10 s/h
Unconstrained delay data 8 and 53 byte	Packet loss < 10^{-6}	Av. delay < 50 ms 90% delay < 100 ms

Source: Reference 3

Whereas today's systems are essentially optimised for speech, future systems will need to be optimised to allow efficient support of these other service types, which will become increasingly important over the coming years.

The 'multi-everything' scenario: Third generation systems will operate in a multi-network, multi-operator, multi-service scenario, supporting what today are perceived as separate cellular and cordless services, as well as services at higher-than-today's data rates. Systems must adapt to support users as their location changes, from a public to private network environment, from one capable of supporting only low rate to high rate transmission, and so on. The ability for multiple networks to geographically coexist, as overlaid public and private networks and as overlaid competing public networks, implies a much greater degree of flexibility and intelligence than today's systems.

Quality of service: The early quality of service of cellular networks, in terms of speech quality, dropped calls etc., was considerably poorer than that of the fixed network; this was a price users were willing to pay for the benefit of mobility. Quality of service has improved dramatically in recent years and there has been a push to improve the quality for future systems closer to the well recognised high standards of the fixed network; whether this is a real market requirement, however, is unclear. One can see that as mobility becomes the assumed norm, and mass mobility markets develop, room will exist for operators to differentiate on the basis of quality of service and associated price. Thus, it is likely that some PTTs will continue to press for future systems to support improved quality of service, whilst others will implement trade-offs in this area.

Multiple cell structures - pico/micro/macro: The concept of picocells emerged from the standardisation of DECT, the digital European cordless telecommunications standard in the 1980s. The low power cordless standards, using omnidirectional antennas, will support ranges well over 100 m, but in a very high usage density environment, e.g. 10,000 E/sq km, the systems are designed to operate in an interference limited rather than range limited mode, resulting in much smaller cells, picocells, of perhaps 10-50 m radius.

The use of microcellular structures has been advocated for many years as a means of providing very high capacity in urban centres without the requirement for large amounts of spectrum. The advent and deployment of low cost cordless standards - CT2 and DECT - for public access, has demonstrated the technical viability of such techniques, and the continued and indeed accelerated fall in cellular technology costs, has allowed the concept to gain widespread acceptance, e.g. for DCS 1800 systems.

For future systems the use of geographically overlaid cell structures is envisaged, regardless of the access scheme used - picocells for private deployments, microcells for pedestrian users, and macrocells accessed by vehicular users - to avoid multiple rapid handovers[1]. The need for handover between cell types is an important resulting requirement.

Infrastructure requirements: Within the public environment, the essential requirement for diversity routeing in the fixed infrastructure for today's CDMA cellular systems (see Section 11.3.1) has been seen as posing an important cost

---

1   The more recent concept of satellite PCN has also now developed similarly, such that handover mechanisms between a satellite cell and macrocells are under examination.

penalty, compared to the conventional (TDMA) hard handoff approach.   Ideally, from the viewpoint of cost and evolution from today's networks, the best solution would be improved quality of service without such a penalty.

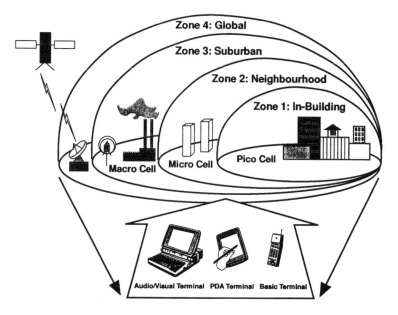

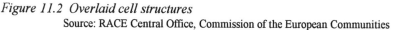

Figure 11.2 Overlaid cell structures
Source: RACE Central Office, Commission of the European Communities

The issue of handover between public and private environments and between different networks will pose major regulatory challenges as well as technological ones; however, the former are not addressed here and whilst the latter will have profound impact upon infrastructure architecture the impact upon the air interface definition will be much less (Chapter 9 discussed the former matter).

Capacity and spectrum usage: Traditionally, system capacity has been seen as a major parameter to be maximised, since spectrum is a scarce resource and, once allocated, an operator's revenue is essentially constrained by his network capacity. Whilst additional smaller cells can be deployed and frequency reuse optimised, the operator will wish to minimise his infrastructure cost per subscriber. (Here again we see a potential benefit in being able to offer flexible quality of service options.) Third generation systems will continue to seek to achieve maximum spectrum efficiency, by the traditional means of efficient modulation and coding techniques and efficient multiple access schemes, as well as by developing new mechanisms for mitigating the effects of co-channel interference, such as described below.

Transmission power: Handheld portable terminals are the major part of the new terminal market today. Such terminals require to maximise battery life - minimisation of transmission power contributes to this goal. This is compatible with an increasing usage of microcell technology, with power control, in urban areas, to maximise system capacity. Public concern over RF safety may also maintain pressure for low transmitter powers. Thus third generation systems must support an efficient low power air interface for handportable terminals, although no doubt higher power options will be available, e.g. for semi-fixed terminals supporting high rate services in picocells.

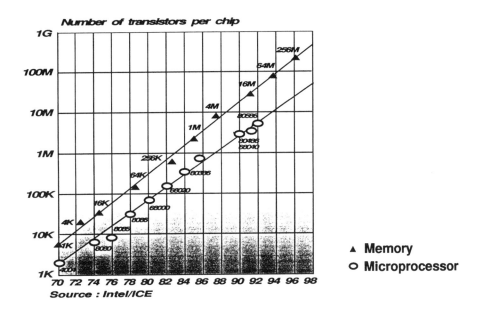

Figure 11.3 Recent advances in silicon memory and processor capability

Implementation technology: The pace of advance of implementation technology must be appreciated in designing future systems. In the 1980s the question of whether (not when) a GSM handportable terminal would be feasible was a major subject of study. The conclusion of the expert study group was that it would be feasible, but even so the pace with which handheld GSM terminals became available and shrank in size and cost surprised many industry observers. Likewise, the progress in silicon capability has been key to allowing the complexities of CDMA systems to be implemented in handheld terminals.

Advances in technology over the next few years will allow the implementation of concepts and algorithms that even today seem incredibly complex; thus third generation standards should be structured to allow the benefits of such foreseen (and unforeseen) advances to maximum effect.

11.3   Application of CDMA and ATDMA to cellular radio

The basic principles of multiple access have been well covered in other references, e.g. [1, 2], and are not replicated here. Instead we build on these to describe the application of such multiple access techniques to cellular radio telecommunication systems. We limit ourselves in this section to a high level explanation of the techniques of CDMA cellular radio and progress in advanced TDMA. It should be appreciated that there are many subtle interrelationships between different aspects of a CDMA cellular radio system design, even more so than with TDMA. For a fuller appreciation of these the reader is referred to References 4, 5 and 6, and to the numerous other papers and patents relating to the IS-95 system and other subsequent systems. As well as addressing Direct Sequence CDMA (DS-CDMA) a brief mention is also made of Frequency Hopping CDMA (FH-CDMA) for completeness.

11.3.1 CDMA cellular - direct sequence

Spectrum spreading: The essential concept of direct sequence spread spectrum is the idea of taking a relatively low-rate, narrowband, digital information stream, and convolving this with a higher-rate, wideband, pseudo-random spreading code to result in a similarly wideband transmitted signal. At the receiver, correlation with an identical code reproduces the original low-rate information sequence, whilst other narrowband interfering signals within the (wide) receiver front end bandwidth are spread across the full code bandwidth, and most of their energy rejected by subsequent narrowband filtering - see Figure 11.4. Signals from other same-system users employ different spreading codes and are thus also re-spread across the full bandwidth and result in a similar, small, measure of interference after filtering, assuming the common case of non-orthogonal codes. If the system uses orthogonal codes then in principle no energy should result from other same-system users within the post-correlation receiver bandwidth. Provided the processing gain G_p (the ratio of the spectrum bandwidths) is sufficiently greater than the ratio of interferer energy to the wanted spread signal then the system can successfully reject own- and foreign-system interferers and operate as intended.

Interferer diversity: In CDMA cellular systems the transmissions from many mobile users within a given cell are spread over the same spectrum, with the processing gain providing a mechanism for discriminating against the unwanted, interfering signals, as described above. The interferers are effectively 'averaged' in a non-orthogonal system, an effect referred to as 'interferer diversity'. This results in a graceful degradation of wanted signal carrier-to-interference ratio, C/I, as the number of users in a cell increases, one factor contributing to the capacity gains offered by CDMA.

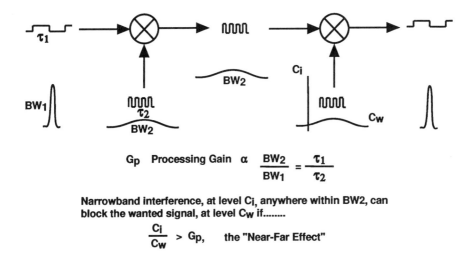

Narrowband interference, at level C_i, anywhere within BW2, can block the wanted signal, at level C_w if........

$$\frac{C_i}{C_w} > G_p, \quad \text{the "Near-Far Effect"}$$

Figure 11.4 The essential concept of direct sequence spread spectrum

Single cell frequency reuse: As a multiple access scheme, CDMA allows the same frequency spectrum to be reused in all cells - the principle of single cell frequency reuse. Thus whereas typical TDMA systems may use a 4- or 7-cell repeat pattern (or other), with different blocks of spectrum in each of the cells, a CDMA system typically can use the same spectrum in all cells. This very high reuse efficiency contributes to high system capacity, since the number of users sharing the same spectrum resource is increased. Further, the traffic level fluctuations of a large number of users will be small relative to the median value, resulting in the need for only a small capacity margin for a specified grade of service (blocking level), further improving average system capacity.

Forward error control: Strong forward error control (FEC) is essential for DS-CDMA systems to achieve good capacity; FEC codes offering some 7dB improvement in fading, with interleaving, are typical. In addition to providing improved performance against interference, strong FEC also allows improved error floors to be achieved - potentially important for some of the advanced low error rate services envisaged for third generation systems. Whereas in TDMA systems a trade-off exists between the use of FEC and performance, in DS-CDMA such a compromise does not exist; i.e. the use of strong FEC with CDMA is entirely beneficial.

Power control: DS-CDMA systems suffer from what has become traditionally known as the 'near-far' problem. This is the situation whereby a large signal from a nearby transmitter exceeds a weaker signal from a more distant transmitter by more than the processing gain, thereby precluding system operation. Various means of avoiding this problem have been proposed, including joint detection and interference cancellation. The approach adopted for IS-95 is the use of fast and accurate control of the transmitter uplink power of all the mobiles in the system. Such power control is both readily implementable and effective.

A further beneficial effect of uplink power control is a perceived reduction in signal fading for slow mobiles. This results in interleaving being needed only to accommodate fading for fast moving mobiles; the interleaving depth can thus be reduced, resulting in a lower service delay.

Downlink power control is also desirable to accommodate the varying levels of interference perceived by the mobile as it moves across the cell. This effect is less severe than the classic 'near-far' problem but inclusion of downlink power control further improves system capacity.

Power control can, of course, also be implemented with TDMA systems, as indeed it is with GSM, although in the latter case it is slow and coarse. However, power control arguably offers greater benefits when matched to CDMA as an access scheme, for reasons described in [7]. With a DS-CDMA system, a finite risk still exists that a single rogue mobile[2] whose power control fails, could transmit at high power, thereby disrupting communications over an entire cell.

Soft handoff: To avoid the near-far problem in regions of overlapping cell site coverage, the mobile transmitter power must always be controlled by the base station requiring the least power from the mobile. Without soft handover, if a mobile roamed into an adjacent cell, that cell's base station would see an increased interference level over which it had no control and which could thus severely degrade system capacity and performance. It is for this reason that soft handover is essential in a DS-CDMA system. In theory, soft handover could be replaced by the use of instantaneous hard handover; however, today's systems actually design in hysteresis, to avoid oscillating handovers. Whilst instantaneous hard handover is not a currently viable option, new handover techniques are an active area of research.

Soft handover has both costs and benefits. Considering firstly the costs, it results in base stations needing to be overequipped with radio modems, compared with a TDMA equivalent, in order to accommodate the many mobiles which find

---

[2]   Whilst such "rogue mobile" arguments are sometimes cited as a weakness of DS-CDMA, one has to ask whether the equivalent GSM rogue mobile, transmitting on all timeslots, may not be equally probable!

themselves simultaneously affiliated to multiple base stations. Additional backhaul traffic capacity, switch processing and diversity combining are all needed to allow the traffic to be routed into and used in the network. The benefits are that it is a means of macro-diversity, thereby offering improved coverage or quality of service. Macro-diversity can be implemented in TDMA systems also, of course, with similar associated costs and benefits, but is optional; in today's DS-CDMA cellular systems it is essential to system operation.

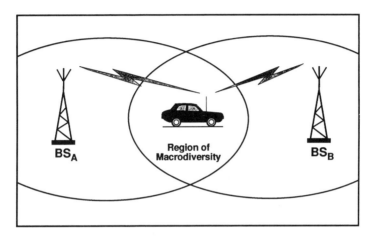

Figure 11.5   *Soft handover: mobile affiliated to two base stations*
The mobile's transmit power is maintained at the lower of the
levels required by the two base stations

For the downlink, separate codes are used for the different base station transmissions to the mobile, which despreads both codes and combines the signals before demodulation. To achieve optimal performance requires very tight synchronisation of the base stations, again with associated cost and operational implications. Such costs must be weighed against the benefits of CDMA; indeed, for third generation systems, approaches which maintain the benefits whilst reducing the costs are an important objective.

Pilot transmissions: Each base station in a DS-CDMA cellular system transmits a 'pilot code' - essentially an unmodulated spread spectrum signal. Correlation of this at the receiver provides the channel estimates needed for cophasing the signal and combining the multipath components.

Voice activity detection: Voice activity detection (VAD) is of course not unique to DS-CDMA; however a larger benefit accrues in a CDMA than in a TDMA system, since 'the law of large numbers' works better - i.e. the wideband CDMA channel can better statistically multiplex the talkspurts to allow improved overall capacity than can a TDMA channel with much smaller numbers.

Sectorisation: Similar arguments also apply as for VAD; sectorisation reduces the number of perceived interferers, and again the statistics work better for the DS-CDMA case.

Advanced detection techniques: Conventional spread spectrum receivers treat unwanted interferers as noise. Better performance can be obtained by exploiting knowledge of their characteristics. This is done in ways which fall broadly into two categories, usually referred to as Interference Cancellation (IC) and Joint Detection (JD).

The essential principle of interference cancellation is that the unwanted signals are demodulated, limited, remodulated and then subtracted from (a delayed version of) the original received signal, in order to improve the apparent C/I for the wanted signal.

Joint detection further exploits knowledge of the codes and channel estimates for the unwanted signals to 'approximate' a *maximum a posteriori* demodulation process - in effect taking a sequence of incoming signals and correlating against a range of possible transmitted bit sequences to identify the most likely transmitted data sequences from all the received signals. This is, in concept, somewhat similar to the role of an equaliser in TDMA systems. Just as in TDMA, suboptimal equalisers are used, instead of the full Viterbi algorithm, to permit much simpler implementation at modest performance reduction, so today's joint detection algorithms similarly use suboptimal approximations.

Today's interference cancellation and joint detection techniques are currently very complex, in terms of required processing power. However, progress in digital silicon density will increasingly facilitate their practical feasibility over the relevant timescales of UMTS, as described earlier.

11.3.2   CDMA cellular - frequency hopping

The GSM standard, as implemented, allows a degree of frequency hopping which an operator may choose to employ. In fact when frequency hopping is implemented (it is optional) it adds an element of orthogonal CDMA to GSM, providing interferer diversity and diversity against frequency selective fading, and hence improved capacity. The use of partly orthogonal hopping also allows 'other than normal' intermediate cell cluster sizes.

FH-CDMA multiple access techniques have been proposed for PCS in North America [8], although they have received less publicity in Europe than DS-CDMA, probably because of the short term US focus on PCS (rather than third generation). Nonetheless, for completeness, FH-CDMA is briefly included in this chapter, as the techniques emerging may find application to third generation systems.

Code orthogonality: FH-CDMA allows the use of orthogonal codes for mobiles affiliated to a common base station, reducing intracell interference compared to a

DS-CDMA system, and thus offering a relative increase in capacity. Orthogonal spreading codes can be used in DS-CDMA, but the time delays associated with multipaths are such as to reduce the potential benefits. Orthogonal codes do provide some benefit and are implemented on the IS-95 downlink, but to work on the uplink would require mobiles to be synchronised to fractions of a chip period, an impractical requirement at present. For FH-CDMA, orthogonal codes can readily be used, since the FH dwell time is typically much larger than a DS chip period, and multipath effects therefore have less effect. Since inbound transmissions to the base station are orthogonal, no near-far problem exists with FH-CDMA and thus power control is not required.

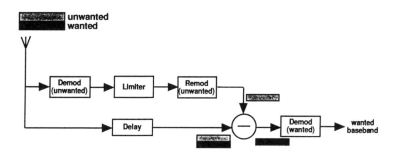

Figure 11.6 The concept of interference cancellation

Interferer diversity and interleaving: Whilst orthogonality offers the prospect of increased capacity, another aspect of FH-CDMA militates in the opposite direction, namely the fact that interleaving of the data over many hops becomes necessary with FH-CDMA, in order to achieve the same degree of interferer diversity as is inherently provided with DS-CDMA. This implies an undesirable trade-off between capacity and service delays - put more simply, it would appear difficult to achieve the necessary short service delays whilst simultaneously supporting high capacity.

Flexibility of service rates: An important requirement for future systems is the requirement to support multiple service rates. FH-CDMA is less well suited to this requirement than DS-CDMA, because orthogonal hopping is not possible at different bit rates and, if high bit rates are used, then equalisation becomes necessary, with its associated overhead of training. High service data rates could be supported using multiple codes but, for FH-CDMA, this would also imply multiple synthesisers and receivers and a major increase in equipment complexity.

11.3.3 Advanced TDMA

TDMA systems were well researched and developed in the 1980s, i.e. the GSM, DAMPS and PDC[3] systems.   Adaptations of traditional TDMA and new techniques are now being developed to extend the system capability of TDMA for current and future systems.   Some of the results are also applicable to CDMA systems and *vice versa* .

Speech coding: Speech coding advances have continued since the definition of the original GSM vocoder, with lower rate and higher quality codecs now emerging.   CELP[4] and VSELP[5] based coders have emerged as effective low rate algorithms, but further techniques will no doubt also emerge.   For future systems, adaptive switching between codec types with cell type/environment is to be anticipated; such adaptation can be done also with CDMA systems. Adaptation of speech coding rate, but using the same algorithm, is included within the IS-95 standard, to maximise capacity (so-called Q-CELP).

Modulation: Constant envelope modulations have been developed in recent years to permit the use of efficient non-linear PAs, although the advent of efficient linear PA technology is now prompting a re-examination of non-constant envelope schemes.   Trade-offs between bandwidth efficiency and C/I performance have resulted in the use of filtered MSK/QPSK based solutions, rather than higher order schemes, for second generation systems; however, the requirement to support a range of services in multiple environments may lead to switching between different modulation schemes for different environments in future systems.

Frame and timing structures: The framing and timing structures of second generation TDMA systems are rigidly fixed - one of the major constraints - but this is an area where gains may be made to support the needs of future systems, in terms of multiple services and multiple data rates.   Much research has been undertaken on multiplexing speech and data services efficiently on a TDMA air interface; development of suitable protocols (e.g. PRMA based) requires innovative algorithms and performance validation by suitable simulations.

Static and dynamic adaptation: Dynamic adaptation, in the limited sense of Dynamic Channel Assignment (DCA), is already used in cordless systems; the extension of such concepts to cellular systems may be expected.   Aside from DCA, however, only relatively recently has more widespread parameter adaptation within TDMA systems begun to be explored.   Current systems are designed to accommodate worst case channel conditions, and thus for much of the time are

---

[3] The Japanese Personal Digital Cellular system (previously known as JDC)

[4] CELP = Codebook Excited Linear Prediction

[5] VSELP = Vector Sum Excited Linear Prediction

effectively 'over-engineered'. Two levels of adaptation are currently the major focus of research efforts - static adaptation of a range of parameters to match the cell type and dynamic adaptation, in-call, of transport parameters within a given cell. Other adaptation processes are also conceivable, e.g. knowledge based systems in the network which learn optimal time-varying adaptation regimes based, for example, on time-of-day or day-of-week traffic patterns.

Static adaptation selects a range of parameters appropriate to the type of cell to which the mobile is affiliated - pico/micro/macro - optimising the air interface, at a coarse level, to the cell type. Typical parameters adapted include basic modulation scheme, carrier separation, transmission symbol rate, slot duration/number of slots per frame, signalling payload and overhead. Whilst the static adaptation of error coding technique and interleaving to the type of service supported is an established technique, extension of the concept in this way is a logical next step.

Dynamic link adaptation aims to accommodate both slow and fast variations in propagation and interference conditions within a cell, to maintain an overall radio bearer target quality of service. A multiple-parameter set (e.g. modulation, error control code and amount of radio resource) is dynamically chosen to maintain the desired service level based on ongoing measurements and feedback via a return channel.

Equalisation and training: Equalisation techniques continue to be researched in two main directions, one being to achieve simpler, but still high performance, implementations. As the required service data rates are increased, so the implied increase in data throughput of the equaliser implies a greater processing load; even with advances in digital silicon technology, improved algorithms offering processing simplifications will be needed.

The other key area of current equaliser research is that of blind equalisation, i.e. equalisation which can operate without the need for a training sequence. This offers the prospect of reducing the signalling overhead and improving the air interface efficiency. Related to blind equalisation are interference mitigation techniques, which may be implemented within the equaliser processing.

Interference mitigation techniques: Techniques for reducing the effect of co-channel interferers, in concept analogous to DS-CDMA interference cancellation, are now emerging for TDMA schemes, e.g. [9]. Initial results indicate that significant improvements can be gained, even at low C/I, with associated increases in equaliser complexity of less than an order of magnitude.

Interference mitigation by means of spatial reuse, sometimes called spatial Division Multiple Access (SDMA), using directional and adaptive base station antennas are also emerging, e.g. [10]. At present these techniques are seen more as an 'appliqué' approach to cellular systems, rather than as an inherent aspect of the system design; however it has to be said, such techniques offer far greater potential when designed-in from a complete system viewpoint, as described in [11].

11.4   Advanced testbed systems

The two most widely publicised third generation air interface developments, in the early 1990s, were undertaken as part of the European Commission RACE[6] programme, within the CODIT and ATDMA projects, between 1992 and 1995 [12, 13].   Here we briefly summarise some of the technical parameters of the physical layer of these air-interfaces and describe one or two key issues. We also briefly discuss other published work on an ATM-compatible CDMA UMTS testbed as well as the FH-CDMA testbed referred to previously. The reader is referred to References 14, 15 and 16 for a fuller understanding of these systems.

11.4.1   The CODIT testbed system

The design philosophy and overall concept of the CODIT system is well described in [14, 15] and the testbed itself in [16]. One base station, one mobile station and a radio network controller are under construction, with well advanced design and build, with demonstrations and laboratory tests anticipated in late 1995.

The CODIT concept was designed to meet the broad requirements of third generation systems as described earlier, in particular offering flexibility in frequency management in a multi-operator, multi-service environment. As such, the system is based on DS-CDMA, with variable spreading factors, multiple chip rates and multiple RF bandwidths.   Chip rates of approximately 1, 5 and 20 Mchips/s are proposed, with each of the offered information rates mapped onto at least one of these three rates (see Figure 11.7). The concept of multiple RF bandwidths allows for flexibility in operator spectrum allocations and in ability to mix-and-match services - e.g. a 10 MHz allocation being used for 2×5 MHz bands or 10×1 MHz bands.   Clearly, inappropriate use of this flexibility, in transmitting high bandwidth services over too low an RF bandwidth, would begin to lose the characteristic benefits of CDMA.

The generic transmission scheme of CODIT comprises signal processing blocks for CRC block coding (frame error detection), convolutional coding, time interleaving, multiplexing and multiframing, frame generation, DS code spreading, modulation, transmitter power control and discontinuous transmission (DTX), linear combination of user channels (on the base station only) and frequency conversion.

Traffic channels and associated or dedicated control channels use independent coding/interleaving branches, which are merged in the multiplexer, allowing both

---

6   RACE = Research and development in Advanced Communication technologies in Europe, an EU sponsored programme of pre-competitive R&D into advanced telecoms, of which mobile communications is a small part. Begun in the mid-80s, RACE is shortly to finish, to be followed by a new programme, ACTS, Advanced Communications Technologies and Services, targetted to achieve technology pull through into the marketplace.

slow and fast logical control channels to be flexibly implemented. In a manner somewhat similar to the static adaptation concept, parameters for all the above functions are controlled by a configuration unit to determine the exact behaviour of each of these blocks. Whilst dynamic adaptation is also in principle possible this was not incorporated in the CODIT testbed.

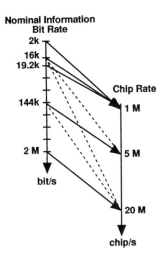

Figure 11.7   Flexible mapping of service bit rates onto chip rates in the CODIT concept

Long pseudorandom spreading codes are used, offering the benefits of a virtually unlimited set of codes, a high degree of flexibility with respect to multiple bit rates and spreading the factors, the ability to operate without inter-channel synchronisation within or between cells and the ability to operate without base station subsystem synchronisation. A loss of capacity results from this non-orthogonality, but this is regained as coherent detection of traffic channels is possible. Techniques such as interference cancellation and joint detection could in future allow higher capacity. Short codes are used for the pilot, synchronisation and random access channels.

Variable bit rate services are implemented by configuring the transmission scheme for the maximum information rate and then simply increasing the spreading factor for periods of lower rate data; the increased processing gain means that the transmission power can then be reduced during these periods. A highly protected, CDMA-multiplexed, Physical Control Channel, PCCH, conveys information on the dynamic information rate to allow configuration of the appropriate receiver processing; this PCCH channel also allows packet services to be supported.

A novel 'compressed mode transmission' handover concept has been developed for CODIT, whereby, whilst handing over between cells on different carriers, the transmitted frame is divided into two halves. In each half the signal is transmitted to each base station at a different frequency at double speed; only half the spreading ratio can be achieved, and this is compensated by doubling the transmit power. The degradation caused by this is expected to be low, assuming that the proportion of time spent in such handover situations is small.

11.4.2 The ATDMA testbed system

The design philosophy and overall concept of the ATDMA system is well described in References 3, 12 and 13. The ATDMA project is developing hardware and software testbeds, the former to validate low level system functionality, the latter to model and validate the performance of higher complexity, deployed network, scenarios. Hardware, which may be flexibly configured, for example, as two base stations, two mobile stations and a network emulator, is being developed. Hardware and software integration is well advanced, with detailed system performance evaluation in hand.

Many of the concepts emerging from the ATDMA project have been alluded to already. The air interface is based on GMSK and Binary/Quaternary Offset QAM, with a choice of carrier bit rates, channel coding and radio bearer types to suit the range of operating environments and applications of UMTS. The transport chain is controlled by a set of functions to set transmit power, link adaptation, ARQ, packet access, channel assignment, handover and admission control. The static adaptation of the air interface with respect to different cell types is summarised in Table 11.2; the process of dynamic adaptation is described in [3]. (It should be noted that the ATDMA burst and frame structures summarised in Table 11.2 have been based upon an assumed 13 kb/s gross rate speech codec - different numbers would result if the base assumption were say a 200 kHz desired carrier spacing.)

Table 11.2   Static adaptation parameters for the ATDMA UMTS testbed system
Source: Reference 3

	*long-***Macro**	*short* **Macro**	**Microcell**	**Picocell**
Modulation	GMSK	Binary Offset QAM		
Carrier symbol rates	360 kbaud	450 kbaud	1800 kbaud	
Minimum carrier separation	276.92 kHz		1107.69 = 4 x 276.92 kHz	
Frame duration	5 ms			
Slots per frame	15	18	72	
Payload	76 symbols	72 or 76 symbols		96 symbols
Training sequence	23 symbols	29 or 33 symbols		15 symbols
Tail bits	8 symbols			6 symbols
Guard time	13 symbols	12 symbols		8 symbols
Slot length	120 symbols	125 symbols		

It should be stressed that projects such as CODIT and ATDMA are not primarily defining proposals for an air interface standard, but rather are developing toolkits and evidence to support third generation standardisation. The system functional model described in [3], for example, is important in this respect as it provides a means whereby a radio access system design may be captured and described in terms essentially independent of the target system network architecture.

11.4.3 The ATM-CDMA testbed system

A new DS-CDMA air interface system for mobile communications is described in [17, 18], which has been designed to exploit the inherent synergies between CDMA and ATM[7]. Whilst the system has been designed to support a range of generic services, such as in Table 1, specific attention has been paid to supporting ATM-based services, to enable optimal integration with future broadband ATM networks, applications and services. Whilst a UMTS air interface will not support high rate ATM, nonetheless flexible support of low to moderate rate services (e.g. up to 256 kb/s) is desirable. Testbed integration of this concept is well advanced, with trials during 1995.

In this system design the Virtual Path Identifier (VPI) and the Virtual Circuit Identifier (VCI) fields of an ATM cell header are mapped onto CDMA codes, whilst other header fields are simplified or removed; this minimises the air interface overhead, compared with simply packaging up entire ATM cells and sending them over a radio interface. The size of the transmitted radio burst, or Air Interface Packet (AIP), is arranged to have an integer relationship with the size of an ATM cell. For high rate services a whole number of AIPs deliver an ATM cell, whilst for low rate services, a whole number of ATM cells are delivered in one AIP. Within the fixed network, knowledge of the CDMA codes allows the full ATM cells to be reconstructed and transmitted onwards to their final destination.

Operation is frequency division duplex, at around 2 GHz, compatible with UMTS allocations. 8 kb/s encoded speech is encoded into 10 ms AIPs and can be carried with a delay of < 35 ms; 5 ms AIPs may also be used and a wide range of data services simultaneously supported. Higher bit rate speech will see smaller service delays. Variable bit rate services, ranging from 8 kb/s to 256 kb/s are supported by varying the CDMA spreading factor, whilst maintaining the spread bandwidth constant at around 5 MHz. Simple BPSK, DPSK and QPSK modulations are used. Pilot transmissions allow coherent downlink demodulation. AIP checksums allow corrupted cells to be identified and discarded and, if appropriate, retransmission requests made. Fast accurate power control allows good C/I performance for slow moving mobiles, with low interleaving depth, contributing to low service delays.

Measurement of packet-by-packet base-station-to-mobile link quality is exploited to allow 'near instantaneous hard handover', so that each uplink AIP can indicate which base station should route it into the fixed network, reducing redundant traffic capacity, a traditional cost associated with today's DS-CDMA

---

7   ATM = Asynchronous Transfer Mode

cellular systems. Requirements for base station synchronisation are minimised by the use of fixed network channel sounding, using ATM cells to provide coarse synchronisation to ~5 ms; finer synchronisation is unnecessary, with the resultant timing uncertainties actually exploited to allow smoothing of the processing loads on time multiplexed equipment.

11.4.4   A FH-CDMA testbed system for PCS

As explained earlier, the FH-CDMA system concepts [8] developed have been targeted at the present day North American PCS opportunity, not third generation. Nonetheless, a brief description of this testbed is included here to give a flavour of FH-CDMA implementation. The main system parameters are given in Table 11.3 below.

Table 11.3   Parameters for the FH-CDMA testbed system
            Source: Reference 8

Frequency Bands	Inbound: 1,899 to 1,929 MHz
	Outbound: 1,949 to 1,979 MHz
Channel spacing	400 kHz
Channel bit rate	500 kb/s
Frequency hopping rate	500 hop/s
Time slots/frame	10
Time slot duration	0.2 ms
Frame duration	2.0 ms
Gross data rate/slot	34 kb/s
Duplex method	FDD
Intracell multiplexing	TDM/FDM (orthogonal SFH)
Intercell multiplexing	SFH/CDMA
Modulation	QPSK
Pulse shaping	Raised cosine, a = 0.5
Receiver gain control	Hard-limiting
Antenna diversity	Pseudo max-ratio
Demodulation	Burst coherent
Transmit power	800 mW
Channel coding type	Rate 1/2, K = 6 convolutional, soft decision
Interleaving span	40 ms
Speech coder	32 kb/s ADPCM

Mobile and base station testbed equipment for this system has been constructed as 19' racks (10' high by 12' deep), with peak transmission powers of 800 mW, suitable for vehicular field tests. Trials have compared achieved BER vs. S/N, against simulations, and have demonstrated the anticipated benefit of diversity against frequency selective fading. A minimum-complexity design philosophy was adopted for the system, to achieve low cost, eliminating the requirement for an equaliser, thereby allowing a hard limiting receiver to be implemented; this in turn resulted in other simplifications, such as allowing cost effective implementation of diversity reception. A block diagram of the PCS transceiver is given in Figure 11.8. A more detailed description of the transceiver can be found in [19].

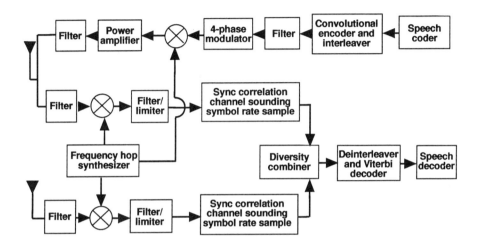

Figure 11.8 Block diagram of the SFH transceiver developed for PCS
    Source: Reference 8

As noted earlier, the FH-CDMA scheme has not been designed with the future multi-service, multi-operator scenario of UMTS in mind. The application of FH to the PCS requirement has resulted in a much simpler implementation than would be possible for third generation systems.

11.5 Conclusions

This chapter has given an overview of some of the emerging air interface techniques likely to contribute to the definition of third generation mobile radio systems - UMTS/FPLMTS - and the design criteria for such systems.

Requirements to support multiple new services, in mixed public/private environments and in a competitive multiple operator environment, will militate toward a flexible and adaptive air interface. Advancing base technology will enable future systems to implement complex algorithms and concepts which today would still seem inappropriate. The future air interface will not however be a simple choice between TDMA and CDMA - even current systems employ elements of both techniques - but more likely a complex hybrid, drawing upon techniques developed from all camps. Whilst spectrum efficiency is important, many other factors, both technical and commercial, will influence the choices.

Current pressures from European cellular operators for evolution from their existing investments, suggest that the GSM system will be stretched considerably, even from today's capabilities and standards, to provide a basic subset of what may have been originally perceived as UMTS features - indeed DCS1800 itself is an illustration of this process already under way. Nonetheless, the rigid basic frame and timeslot structures of GSM impose some fundamental limits on the data rate and thus service capabilities of the system. At some stage, especially with the advent of higher data rate services in fixed telecommunications networks, we can envisage potential markets emerging that will require a significantly more flexible air interface, embodying some of the concepts described herein. It is to the definition of these standards that significant engineering efforts will be applied over the next few years.

Acknowledgements

Some of the work described in this chapter (the ATDMA and CODIT projects) has been partially funded by the European Community in the RACE Programme (Research and development into Advanced Communications technologies for Europe).

In the context of the RACE 2084 (ATDMA) work, the author would like to acknowledge the contribution of colleagues from Siemens AG, Roke Manor Research, ALCATEL Radiotelephone, ALCATEL Standard Electrica S.A., ALCATEL Italia S.P.A., Universidad Politécnica de Cataluña, Télécom Paris, France Telecom CNET, ESG Elekronik-System-Gesellschaft mbH, Fondazione Ugo Bordoni, The University of Strathclyde, DeTeMobil, Nokia Mobile Phones and Nokia Research Centre.

In the context of the ATM-CDMA air interface work, the author would like to acknowledge the contribution of colleagues from Roke Manor Research, Siemens and GPT.

In the context of the RACE 2020 (CODIT) work, the author would like to acknowledge the assistance of partners within that project in providing copies of relevant recent publications referenced herein.

Whilst the views expressed in this paper have been shaped by interaction with industry colleagues from many organisations over recent years, they should be interpreted as a purely personal view, rather than that of any organisation.

References

1   Balston, D.M. and Macario, R.C.V., *'Cellular Radio Systems'*, ISBN 0-89006-646-9, Artech House, 1993

2   Tuttlebee, W.H.W. *'Cordless Telecommunications Worldwide'*, ISBN 3-540-19970-5, Springer Verlag, 1995 (see also *'Cordless Telecommunications in Europe'*, Tuttlebee, W.H.W. ISBN 3-540-19633-1, Springer Verlag (1991)

3   Urie, A., Streeton, M., Mourot, C. 'An Advanced TDMA Mobile Access System for UMTS', *IEEE Personal Communications Magazine*, vol 2, no 1, pp 38 - 47, February 1995

4   'On the System Design Aspects of CDMA applied to Digital Cellular and Personal Communications Networks', A Salmasi, KS Gilhousen, *Proc 41st IEEE Conference on Vehicular Technology*, St Louis, Missouri, May 1991

5   *'Mobile Station - Base Station Compatibility Standard for Dual Mode Wideband Spread Spectrum Cellular System'*, TIA/EIA Interim Standard IS-95, July 1993

6   Numerous papers on the topic of CDMA, some relevant to mobile systems, may be found in the *IEEE Journal on Selected Areas in Communications*, two part special issue on 'Code Division Multiple Access Networks', v12, no 4 and 5, May and July 1994.

7   Hulbert,A.P. 'Myths and Realities of Power Control', IEE *Colloquium on 'Spread Spectrum Techniques for Radio Communications Systems'*, London, April 1993

8   Rasky, P.D. Chiasson, G.M. Borth D.E. and Peterson R.L. 'Slow Frequency-Hop TDMA/CDMA for Microcellular Personal Communications', *IEEE Personal Communications*, pp 26-35, vol 1, no 2, 1994

9   'A Technique for Co-channel Interference Suppression in TDMA Mobile Wales, S.W. Radio Systems', *IEE Proceedings Communications*, pp106-114, vol 142, April 1995.
       An abridged version of this paper was presented at the IEE Colloquium on 'Mobile Communications towards the Year 2000', London, 17th October 1994

10   'Adaptive Antennas for Frequency Reuse in Every Cell of a GSM Network', M Wells, presented at the IEE Colloquium on 'Mobile Communications towards the Year 2000', London, 17th October 1994

11   'Evolutionary Path to 100% Spectrum Re-use for GSM Cellular Radio Applications using Adaptive Antennas', M Wells, Roke Manor Research, submitted to *IEE Proceedings Communications*, 1995

12   IEEE Personal Communications, Vol 2, no 1, February 1995
This complete issue of the journal was devoted to 'The European Path towards UMTS', and contains several articles of relevance to the subject of this chapter, including descriptions of the ATDMA and CODIT radio access designs, presented in a very readable manner.

13   Grillo, D., Metzner, N. Murray, E.D. 'Testbeds for Assessing the Performance of a TDMA-Based Radio Access Design for UMTS', *IEEE Personal Communications*, pp 36-45, Vol 2, no 2, April 1995

14   Baier, A, Fiebig, U-C, Granzow, W., Koch, W., Teder, P., Thielecke, J. 'Design Study for a CDMA-based Third Generation Mobile Radio System', *IEEE Journal of Selected Areas in Communications*, vol 12, pp733-743, May 1994

15   Baier, A.'Open Multi-rate Radio Interface Architecture bad on CDMA', *Proc. 2nd IEEE International Conference on Universal Personal Communications*, pp 985-989, Ottawa, Canada, 1993

16   Andermo, P.G., Larsson, G. 'Code Division Testbed, CODIT', *Proc. 2nd IEEE International Conference on Universal Personal Communications*, pp 397-401, Ottawa, Canada, 1993

17   Chandler, D.P., Hulbert, A.P. McTiffin M.J. 'An ATM-CDMA Air Interface for Mobile Personal Communications', , paper E1.3, Personal Indoor and Mobile Radio Communications Conference, PIMRC, The Netherlands, September 1994

18   McTiffin, M.J., Hulbert, A.P., Ketseoglou, T.J., Heimsch W. and Crisp, G.'Mobile Access to an ATM Network Using a CDMA Air Interface', *IEEE Journal of Selected Areas in Communications*, 1995.

19   Rasky, P.D., Chiasson, G.M., Borth, D.E. 'An Experimental Slow Frequency-Hopped Personal Communication System for the Proposed US 1850-1990 MHz Band', *Proc. 2nd IEEE International Conference on Universal Personal Communications*, pp 931-935, Ottawa, Canada, 1993

Chapter 12

GSM data, telematic and supplementary services

Ian Harris

Introduction

Whilst communications traffic carried on the GSM cellular network is predominantly voice, provision is made to support a number of data, telematic and supplementary data related services. Only those services which are seen by most GSM network operators to be the most important are discussed here.

Since the early 1980s data services have been available on the UK TACS (Total Access Communication System) analogue cellular network in which the number of conventional data users (excluding facsimile etc.) is estimated to be 1% to 2% of the total number of mobile subscribers. Some bold estimates put the projected figure for the GSM network at 5% for the same type of services. It is likely that standardisation will encourage the use of data on GSM whereas in the UK TACS network there was no standardisation agreed for 'data' between the network operators despite efforts to do so.

One GSM supplementary service, in particular, is the Short Message Service (SMS). This is currently attracting considerable interest from potential users. For certain applications, the short message service is providing an attractive alternative to the more conventional circuit switched GSM data services.

12.1 Cellular radio transmission impairments

Radio transmission impairments in the cellular radio environment are unlike those in a conventional PSTN environment or any other fixed network environment.

Obstacles such as buildings or vehicles in the radio path can put the receiver in a radio shadow or cause the radio signals to be reflected such that a cancellation of signal levels occurs at the receiver. The former impairment is known as 'shadowing' and the latter is known as 'multipath' or 'fading'.

A received radio signal has an inherent noise level at the receiver. This noise may be due to the radio environment, or noise within the receiver circuitry itself. A typical FM radio receiver may have a noise level of about -120 dBm. The wanted received signal level in a cellular environment may be as low as -110 dBm.

The effect of multipath and shadowing is to cause the wanted signal to momentarily fall below the noise threshold . This is illustrated in Figure 12.1.

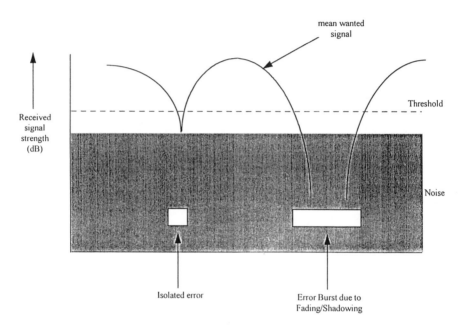

Figure 12.1   Effect of fading and shadowing

In severe conditions the effect may be prolonged and after several seconds the network or the cellular phone will in all probability terminate the call. If the obstacle, the transmitter or the receiver are moving the problem is compounded. It is fairly common within a cell boundary for the receiver to experience bit error rates of the order of 2% (i.e. 1 in 50 bits may be in error). By contrast, PSTN bit error rates are rarely worse than 1 in 10^{-5}.

12.2   Error protection techniques

There are two techniques used in GSM to combat the transmission impairments mentioned above:
- Protocols
- Forward error correction

12.2.1 Protocols

In many fixed network data communications environments, the integrity of transmitted data is usually safeguarded by a layer 2 protocol. Such protocols have the ability through a Cyclic Redundancy Check (CRC) code appended to the transmitted data to detect the presence of data errors at the receiver and to automatically request re-transmission of erroneous data. This retransmission mechanism is known as 'ARQ'.

An example of an ARQ protocol is the High Level Data Link Control (HDLC) layer 2 protocol standard or CCITT X25. Figure 12.2 illustrates the principle of such a protocol. The data to be transmitted is divided up into blocks, each block having a Cyclic Redundancy Checksum (CRC), sometimes also referred to as the Frame Check Sequence (FCS), which is the result of a mathematical computation of all the preceding data in the block. The header of each block contains supervisory information and in particular a send and receive block sequence number, which indicates the number of the last sent and last received block within the modulus window of the block sequence numbers.

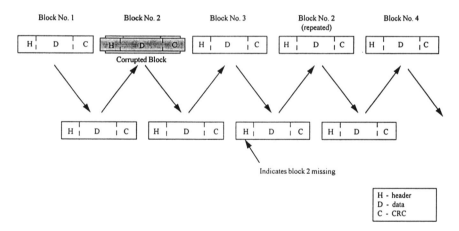

Figure 12.2 Example of ARQ in a layer 2 protocol

If an erroneous block is received, the receiver informs the transmitter which then retransmits the block at the earliest opportunity.

In severe error conditions, it is possible for data transmission to cease through an effect known as 'windowing'. Windowing will occur when the outstanding erroneous blocks are unable to be successfully transmitted within the block sequence number modulus window. If the error rates remain high during the error recovery it is likely that the layer 2 protocol will collapse as neither end of the link will be able to ascertain the state of the opposite end.

In GSM, a protocol very similar to HDLC is used and is known as RLP. This is discussed in more detail later.

To improve the performance of a layer 2 protocol in a noise prone environment a technique known as Forward Error Correction (FEC) may be used.

12.2.2 Forward error correction and interleaving

The basic principle of forward error correction is to add additional information to the source data so that in the event of bit errors occurring during transmission, the receiving entity is able to use the additional information to correct the bit errors without the need for ARQ. Clearly there is a limit to the number of bit errors which may be corrected in this way, and so in practice FEC merely reduces the frequency of ARQ occurrences.

There are a number of different FEC codes, e.g. symbol block codes, convolutional codes and binary block codes. The Reed Solomon (RS) codes are probably the best known symbol block codes and tend to be more effective in dealing with burst errors rather than random errors. Convolutional codes tend to be more complex to implement. Binary block codes can be very effective in correcting for a large number of random bit errors and are relatively simple to implement. A compromise between implementation complexity, code performance in the given environment and redundancy overhead often has to be made when selecting an FEC code. For GSM, convolutional coding was chosen. The degree of redundancy is expressed as two numbers. In one example of a convolutional code for a GSM traffic channel for user data there are a total of 114 bits of which 61 are source data and the remainder the overhead for FEC.

In order to strengthen the effectiveness of an FEC code a technique known as 'interleaving' is often used. Interleaving effectively distributes the effect of burst errors by reorganising the sequence of FEC encoded data bits so that the FEC algorithm may be applied to bits which have an interval greater than the predicted burst error period. The interval is known as the interleaving depth. Increasing the interleaving depth has the disadvantage of increasing the transmission delay. Figure 12.3 illustrates the principle of interleaving. A burst error occurring during a non interleaved bit stream where the error burst period was greater than the period of the block itself would cause the entire block of data to be erroneous. However if all the first bits of all the blocks to be transmitted were gathered together and sent as a block and the same was done for all other bits in each original block then the same burst error would only remove 1 bit from each of the original blocks. In this way since the original block will already have additional FEC data, the receiving entity will be able to correct for a single bit error in each of the original blocks.

There is of course a limit to the number of bit errors which can be corrected with FEC algorithms. The performance of interleaved FEC codes used for GSM is discussed later.

The effect of high bit error rates in the GSM radio environment is that the supervisory/control channels experience delays due to their inherent re-transmission mechanism. Traffic channels which are used for data (and facsimile) may experience corrupted information or delayed information depending upon whether an FEC only service or an ARQ service using RLP is being used. These services are discussed in more detail later.

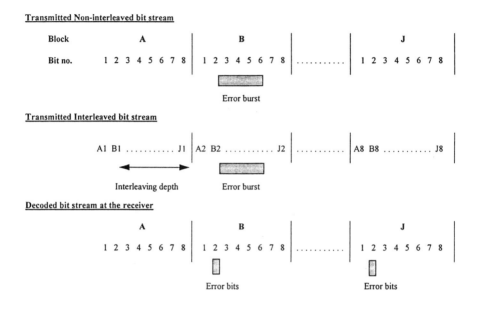

Figure 12.3 Interleaving principles

12.3 Data services

Although certain Private Mobile Radio (PMR) systems have provided PSTN and dedicated data network interworking for some years, their growth and success has been limited for a variety of technical and commercial reasons such as limited radio coverage, application specific products and the inability to support full duplex communications. By contrast, the GSM cellular radio network offers virtual nation-wide coverage and each base station has independent transmit and receive channels which enable simultaneous both way (full duplex) communications. Additionally, GSM provides users with the ability to roam to other countries where because of ETSI standardisation, services will be supported in much the same way.

Figure 12.4 provides an overview to how 'data' is supported in a GSM environment. Up to 8 or 16 simultaneous traffic channels may be contained within a single radio frequency carrier through the use of Time Division Multiple Access (TDMA). Where 8 such channels are provided within a single RF carrier each channel is known as a Full Rate Channel (FRC). Where 16 such channels are provided each channel is known as a Half Rate Channel (HRC). The half rate channel was developed initially for speech in order to conserve bandwidth although some degradation in quality is inevitable. Half rate channels are also used for user data rates of 4.8 kb/s and below, which do not necessarily require the capacity of a full rate channel.

It is possible that operators may wish to apply differential tariffs for the use of full and half rate channels.

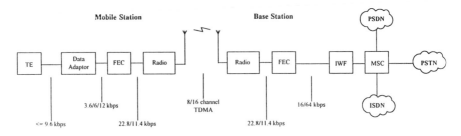

Figure 12.4 GSM data schematic

Each radio channel comprises a mixture of traffic and control TDMA time slots. In one example where a speech or data call is in progress, 24 time slots are assigned as Traffic Channels (TCH) and 2 are assigned as Slow Associated Control Channels (SACCH).

Further information on the channel structure used in GSM can be found in earlier chapters in this book.

At present, a traffic channel assigned to any one user may only be used for one purpose at a time (i.e. speech or data) although there may of course be a mixture of speech, data and telematic services within each RF carrier.

GSM supports data services in a number of different ways and in so doing, provides the user with a choice of data rates, error protection methods, terminal types, fixed network interworking etc.

Unlike the UK TACS analogue network there is no necessity for modems in a GSM environment other than those situated in the Inter-Working Function (IWF) which provides interconnection and interworking with the PSTN. In addition to the PSTN the IWF provides interworking with packet switched data networks (PSDNs) and the ISDN. In this way the cellular environment is isolated from the multiplicity of fixed networks and their equally numerous standards.

A data adapter in the mobile station converts the digital user data rates provided by conventional DTEs to 3.6, 6, or 12 kb/s. FEC and interleaving is performed on each of these rates to give either 22.8 kb/s for a full rate channel or 11.4 kb/s for a half rate channel. The inverse functionality takes place at the base station/MSC except that in general, the data rate between the base station and the MSC is either 16 kb/s or 64 kb/s depending upon the type of connection between them.

Data rate conversion in the data adapter and at various stages in the network is achieved through the use of CCITT V110 rate adapters RA0, RA1, and RA2.

The RA0 function converts asynchronous start-stop data rates to their equivalent synchronous rate by stripping off the start and stop bits.

The RA1 function converts the various synchronous rates to 16 kb/s by multiplexing the data bits and adding synchronisation and control information.

The RA2 function converts the 16 kb/s rate to 64 kb/s. The RA2 function will be typically found at the base station where the interconnection to the IWF is 64 kb/s.

A special modified RA1 rate adapter known as RA1' (RA1 prime) has been defined by GSM in order to provide the correct data rate to the FEC function for maximum utilisation of the available traffic channel capacity. Figure 12.5 shows the format of the 80 bit RA1 frame compared with the 60 bit RA1' frame. The 17 frame synchronisation bits and the 3 'E' bits which are normally used to define the data transfer rate have been removed from the RA1 80 bit frame. Synchronisation for GSM is achieved by other means and the data transfer rate is indicated in the BCIE. User data is carried in the 'D' bits and status such as flow control is carried in the 'S' bits. By presenting the FEC encoder with a 60 bit frame every 5 ms for a full rate channel and every 10 ms for a half rate channel the data input rates of 12 kb/s and 6 kb/s required by the FEC encoder are satisfied. For the lower user data rates where an FEC encoder input rate of 3.6 kb/s is required, the 60 bit frame is further modified to a 36 bit frame by reducing the number of 'D' bits to 24. The 36 bit frame is then presented to the FEC encoder every 10 ms. Figure 12.6 shows how the RA1 prime function may be used in the mobile station data adapter.

When a call set up is requested from or to a GSM mobile phone, certain resources within the GSM network have to be assigned for the duration of the call according to the nature of the call. The resources required are conveyed in Bearer Capability Information Elements (BCIE). The BCIE comprises a number of octets and must be set by the mobile station or the IWF as part of the signalling requirements at the start of the call. Typical examples of the content of the BCIE related to data calls are data rate, parity, flow control, and whether V42 error correction is needed between the IWF and the PSTN modems.

Data calls originating from the PSTN pose a particular problem in that there is no signalling associated with the data devices other than that generated by the modem in the IWF. In such a case the PSTN caller cannot send the BCIE to the IWF as there is no mechanism to do so. The IWF must deduce its own BCIE from the call itself. When the mobile answers the call it will send its own BCIE which the IWF will check to see if it is compatible with the data call being offered. Whether the call is allowed to proceed if there is not a perfect match is an implementation matter in the IWF. Clearly for dedicated applications, there is a better probability that the BCIEs will match. However for a more open system such as a mobile receiving calls from various PSTN sources where there is no prior knowledge or arrangement between the two users, it is likely that the call will fail.

OCTET NO.	BIT NUMBER							
	1	2	3	4	5	6	7	8
1	0	0	0	0	0	0	0	0
2	1	D1	D2	D3	D4	D5	D6	S1
3	1	D7	D8	D9	D10	D11	D12	X
4	1	D13	D14	D15	D16	D17	D18	S3
5	1	D19	D20	D21	D22	D23	D24	S4
6	1	E1	E2	E3	E4	E5	E6	E7
7	1	D25	D26	D27	D28	D29	D30	S6
8	1	D31	D32	D33	D34	D35	D36	X
9	1	D37	D38	D39	D40	D41	D42	S8
10	1	D43	D44	D45	D46	D47	D48	S9

Figure 12.5   CCITT V110 frames

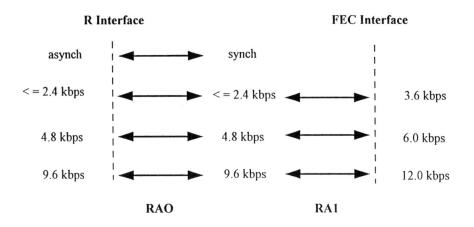

Figure 12.6   Rate adaptation

Bearer services

GSM provides a choice of two bearer services for transporting user data between the mobile station and the IWF:

- The transparent bearer service
- The nontransparent bearer service

The transparent bearer service

This service provides protection for both synchronous and asynchronous user data on the traffic channel through the use of FEC and interleaving. Figure 12.7 shows the characteristics of the FEC codes used in GSM and the predicted performance within a cell boundary (i.e. over 90% of the cell area for 90% of the time) for various user data rates. User data integrity cannot be guaranteed and in addition the service has a fixed end to end delay of several hundreds of milliseconds and so any application or layer 2 protocol protection provided by the user must take this into account.

The nontransparent bearer service

This service overlays the transparent bearer with an integral layer 2 ARQ protocol known as the Radio Link Protocol (RLP) which will virtually guarantee that user data will be transferred to the application layer error free. However due to the effect of ARQ, it is possible that data transfer delays will occur and may be up to several seconds. It is therefore important that the application is able to tolerate such delays and that flow control mechanisms are implemented; otherwise data may be lost due to buffer overflow or calls may be terminated due to timeout in the application.

RLP is based on the IS4335 balanced class of procedures, more commonly known as HDLC.

User Data Rate	Channel	Convolutional Code Redundancy	Interleaving Block Length	Residual BER (after FEC & interleaving)
9.6 kbps	Full rate	61/114 = 0.53	95 ms	$10^{-2}/10^{-3}$
4.8 kbps	Full rate	1/3	95 ms	$10^{-4}/10^{-5}$
4.8 kbps	Half rate	61/114	190 ms	$10^{-2}/10^{-3}$
<2.4 kbps	Full rate	1/6	40 ms	better than 10^{-6}
<2.4 kbps	Half rate	1/3	190 ms	$10^{-4}/10^{-5}$

Figure 12.7 Predicted GSM data error rates

The layer 2 RLP frame comprises 240 bits, of which 192 bits are actual user data, 8 bits are for status (e.g. CCITT V24 control lines) and the remainder are for the RLP header and the FCS.

Figure 12.8 shows the frame structure and content of the frame header. The RLP frame is presented to the FEC encoder every 20 ms in the case of a full rate channel and every 40 ms in the case of a half rate channel thus satisfying the FEC encoder rates of 12 kb/s and 6 kb/s respectively.

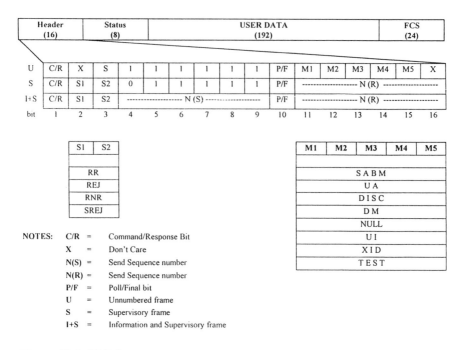

Figure 12.8 RLP frame structure

Whilst the procedural elements of HDLC have been retained in RLP, an enhancement has been made to combine supervisory frames (S frames) with user data frames (I frames), a technique known as piggybacking, which provides a more efficient error recovery mechanism in the event of re-transmission. The IS4335 selective reject error recovery supervisory command is provided in addition to reject. Selective reject allows a single erroneous I frame to be re-transmitted at the next send opportunity whereas reject provides a 'go back N frames' mechanism. Both these features provide a greater efficiency for error recovery in the event of an ARQ. Because the round-trip delay of the underlying transparent bearer can be of the order of a few hundred milliseconds a window size of 64 is necessary to avoid 'windowing'. A 24 bit FCS is necessary to ensure that the probability of the CRC failing to detect an error in the high error rate environment of cellular radio is negligible or comparable to the performance of a 16 bit CRC in common use in fixed networks such as X25.

It is beyond the scope of this section to explain all of the elements of the RLP frame. They are essentially the same as those used in HDLC and a detailed description of them will be found in References 1, 2 and 3.

12.4   The GSM mobile phone

A mobile phone is defined by GSM to be one of three types:

- An MT0 is a fully integrated mobile station with no external interfaces other than possibly a man machine interface. Such a station may be a PC with a built in cellular phone
- An MT1 provides an 'S' interface to allow the attachment of ISDN computer terminals known as TE1s
- An MT2 provides an 'R' interface to allow the attachment of 'V' (asynchronous or synchronous) or 'X' series terminals known as TE2s. Call establishment procedures in accordance with CCITT V25 bis, or X21, are also specified using V24 or V11 interface circuits respectively

These mobile phone types are shown in Figure 12.9.

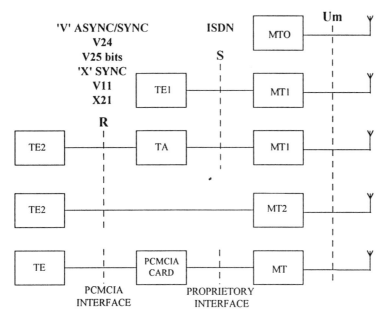

Figure 12.9 Mobile station data reference points

A fairly recent development are PCMCIA (Personal Computer Memory Card International Association) cards for GSM mobile phones. Although they do not strictly comply with any of the above configurations they do provide a convenient way (although currently quite costly) to connect a PC directly to a mobile phone provided the PC has a PCMCIA slot. Most mobile manufacturers offering PCMCIA cards have put the 'data' functionality in the PCMCIA card so as not to increase the weight , cost, size and battery consumption for those users who do not require 'data'. Whilst this argument has some merit it is more concerned with the characteristics of mobile phone design and unfortunately has resulted in a move away from the benefits of standardisation since one manufacturer's mobile PCMCIA card and its associated mobile phone interface is incompatible with another's.

Another recent development is the adoption by some manufacturers of V25 ter for data call control. ETSI has not yet incorporated V25 ter into its standards.

12.5   Future developments

Also, at the time of writing, GSM is planning two major enhancements to its data services:

- Higher user data rates
- Packet radio service

Higher user data rates
It is intended to provide support for higher user data rates by extending the concept of half and full rate channels to multiple channel assignment up to at least 64 kb/s. The on air user rate is currently limited to 16 kb/s which is a serious limitation for ISDN users and applications such as video conferencing and image transmission. The proposal is not without potential problems however. Not only is there a cost implication for users because of the allocation of a large resource to a single user but also technical and operational problems of guaranteeing continuation of resource allocation in the event of a handover to another cell or MSC. There is also a cost implication for network operators since in order to offer the service, additional capacity may be required on some cells for a service which is rarely used.

Packet radio service
Many data applications are more suited to the use of a virtual circuit rather than a circuit switched call. This is because the data transfer duration for many applications is small or bursty. The use of a circuit switched call invariably means that for short duration data transfer applications it often takes much longer to establish the call than it does to send the data. Circuit switched calls are also an inefficient use of the scarce radio resource for the majority of data calls during which there are often long idle periods.

GSM is currently specifying a GSM packet radio service (currently referred to as GPRS) which is intended to support both bursty frequent data applications and infrequent lengthy data applications.

It will be possible to interwork with packet switched networks such as X25 using the same access procedures as those used in the fixed network. The mobile phone will have an X121 address like any other terminal in the fixed network as well as its normal telephone number. The use of TCP/IP is also planned. In terms of timescales, it is possible that GPRS could be in service in some form by 1997. Much will depend on the implementation complexity and changes required to the existing cellular infrastructure. A key consideration in the design is to ensure that any changes to base stations are minimised.. This is because most network operators have typically thousands of base stations currently operational and the cost of modifying them could be considerable and has to be justified commercially.

12.6   The facsimile teleservice

There are literally millions of facsimile machines in use throughout the world, the majority of which are analogue group 3 machines which use modem technology and operate in accordance with CCITT standards. CCITT T4 defines how information is encoded into a facsimile page and CCITT T30 defines how that encoded information is transferred between two facsimile machines.

Two facsimile teleservices are provided by GSM:

- The transparent facsimile teleservice
- The nontransparent facsimile teleservice

The transparent facsimile teleservice carries the T30 protocol and the T4 encoded information on the forward error corrected and interleaved transparent bearer described earlier. This service emulates as far as possible the characteristics of the PSTN with the exception that it has a fixed delay of several hundreds of milliseconds.   The nontransparent facsimile teleservice carries the T30 protocol and T4 encoded information in RLP frames using the nontransparent bearer described earlier. In so doing, a degree of complexity is introduced since the T30 protocol is unable to cope with variable and sometimes large delays which may be caused by the ARQ mechanism of RLP. Facsimile machines cannot be flow controlled for very long and therefore buffering has to be provided.

Whilst there are a number of users of GSM facsimile services today there is insufficient information concerning their performance. However, the main characteristic differences between the two services from a technical viewpoint are as follows:

- The nontransparent facsimile service will in all probability be prone to premature call termination because of time-outs in the facsimile apparatus caused by RLP ARQs although data quality should be very good. The transparent facsimile service will from time to time have the performance characteristics of a poor quality PSTN line but should not prematurely disconnect.

- The majority of GSM network operators have chosen to support the transparent rather than the nontransparent facsimile service.

The nontransparent facsimile service is not described in any more detail here but further reading can be found in Reference 6.

12.6.1 Connecting facsimile apparatus to the mobile phone

Figure 12.10 illustrates a number of ways in which a group 3 facsimile machine may be connected to a GSM phone.

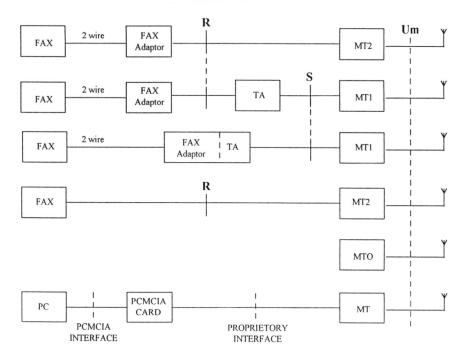

Figure 12.10   Mobile station fax configurations

Although the speech coders used in the mobile phone and the GSM network emulate a 3.1 kHz PSTN path they are unable to operate satisfactorily when presented with CCITT V29 or V27 ter modem modulation used in the facsimile message phase. It is therefore necessary in both the mobile phone and in the IWF to incorporate fax adapters which effectively terminate the facsimile modem functionality. The mobile fax adapter also has to incorporate some of the properties of a PABX in order to generate ringing current for incoming calls and to convert DTMF or loop disconnect dialling from the facsimile machine into V25 bis auto calling procedures for outgoing calls.

Operational problems may exist for some facsimile machines having a telephone handset since that handset cannot be used for speech but may have to be used to initiate or answer a facsimile call.

At the time of writing there are only two known suppliers of a stand alone fax adapter; both adapters are expensive. There is also very little indication that facsimile machine manufacturers are as yet intending to develop a facsimile machine with an 'R' interface which would obviate the need for the fax adapter at the mobile phone.

Ironically it is the non standardised PCMCIA card which has emerged as the market leader in providing facsimile capability for the mobile phone.

12.6.2 The CCITT T30 protocol

It is probably necessary to understand a little about the operation of CCITT T4 and T30 in order to appreciate the complexity in supporting facsimile in a digital GSM environment.

The CCITT T30 protocol has a control phase and a message phase.

The control phase operates at 300 bps half duplex and is responsible for negotiating the capabilities between facsimile machines and supervising the transmission of pages of information. A 'command/response' repeat mechanism is built into the control phase procedures.

The message phase operates at either 9600, 7200, 4800 or 2400 bps half duplex and is responsible for transmitting pages of facsimile information encoded according to CCITT T4. A T4 encoded page of A4 text may comprise typically 40,000 bits.

At the start of a facsimile call, the answering facsimile machine offers a set of capabilities such as transmission speed, from which the calling facsimile machine must select a compatible set or subset. The calling facsimile machine then sends a modem training sequence according to CCITT V29 or V27 ter followed by a sequence of binary zeros at message phase speed known as the Training Check Flag (TCF).

The purpose of the TCF is to enable the receiving facsimile machine to check the quality of the connection. The receiving facsimile machine indicates its acceptance or rejection of the quality by sending a control phase response. If the receiving facsimile machine finds the TCF acceptable then the sending facsimile machine sends the T4 encoded page of information at message phase speed. If the receiving facsimile machine finds the TCF unacceptable then the sending facsimile machine will send a new modem training and TCF sequence at the next lowest message phase speed.

A typical T30 procedure between two group 3 facsimile machines is shown in Figure 12.11.

T30 also provides an optional error correction mode. If both the sending and receiving facsimile machines declare and select the error correction mode during the negotiation at the start of the call, then each page of information to be transmitted is split up into partial pages. Each partial page is subdivided into a number of frames (usually 256) and each frame has its own CRC checksum. In the event of an error being detected in any one or number of frames, the sending facsimile machine is requested to re-transmit those frames which were in error. The error correction process can repeat itself a number of times thereby reducing the number of outstanding frames that are in error on each repeat before proceeding to the next partial page.

12.6.3 Support of the CCITT T30 protocol in GSM transparent facsimile teleservice

The fax adapters have to terminate the facsimile machine modem functionality and elements of the T30 protocol and to signal to each other when to change between the T30 binary control phase (BCS) and the message phase (MSG). This has a significant impact on the tight timing controls specified in T30 which work satisfactorily over a PSTN connection where delays are minimal. Although the T30 protocol is half duplex, the transparent facsimile service protocol is full duplex. This ensures that the mobile fax adapter and the fax adapter in the IWF are always aware of each other's state.

In order to ensure that the data rate across the air interface does not have to continually change between the T30 control phase speed (300 bps) and the T30 message phase speed (9600, 4800, 2400 bps) the BCS frames are multiplexed up either 8, 16 or 32 times so that they can be sent using the message phase speed.

Transfer of BCS frames and facsimile TCF and message phase data between the mobile and the IWF is achieved through the use of a synchronous 64 bit frame structure shown in Figure 12.12.

The link between the fax adapters in the IWF and the mobile phone can be in one of five states which indicate the state of the local facsimile machine connected to that fax adaptor. These states are   IDLE, BCS TRAnsmit, BCS RECeive, MeSsaGe TRAnsmit, MeSsaGe RECeive.

The IDLE state essentially provides a means of byte synchronisation.

The BCS REC state indicates that the fax adapter is receiving T30 BCS data from its local fax machine.

The BCS TRA state indicates that the fax adapter is sending T30 BCS data to its local fax machine.

The MSG REC state indicates that the fax adapter is receiving T4 encoded data from its local fax machine.

The MSG TRA state indicates that the fax adapter is sending T4 encoded data to its local fax machine.

The control of the modem in the fax adapters in the IWF or in the mobile to switch between T30 BCS and message phase speed is achieved by detecting the appropriate link state.

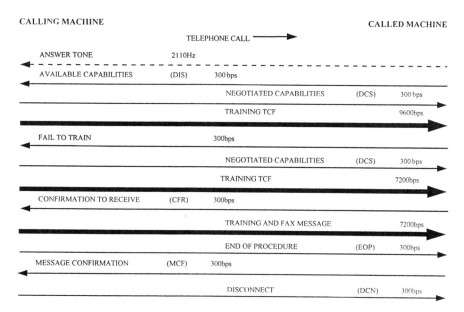

Figure 12.11 Example of group 3 fax CCITT T30 procedures

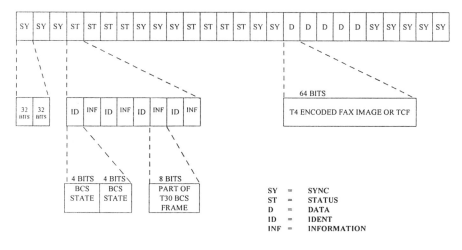

Figure 12.12 Transfer of facsimile information between MS & IWF

The idle state is indicated by sending sync frames when there is no other information to send in a particular direction. Each 64 bit sync frame is split into two sequential 32 bit codewords, the second one being the complement of the first. This allows for up to 3 bit errors in each codeword.

When the BCS state is being indicated status frames are sent. Each 64 bit 'status' frame is split up into alternate 'information' and 'ident' 8 bit fields. The information octets convey the BCS frame information as specified in T30 and the ident octets convey the BCS state repeated twice within the same octet. The repetition of the ident octets allows the receiving entity to 'vote' in order to ascertain the correct link state in the event of bit errors across the link. The T30 BCS information encoded in the information octets have their own CRC protection as defined in T30. The T30 BCS frames comprise more than one octet and are presented to the 64 bit frame bit stream at 300 bps. It is therefore necessary to generate 1, 2 or 4 64 bit status frames every 8 bits of 300 bps T30 BCS data depending upon whether the air interface rate is 2400, 4800 or 9600 bps.

The message state is indicated when TCF or message phase data transfer is in progress. In this case, the information octets in the status frame are coded with a constant as it has no meaning. The T4 encoded facsimile image or the TCF is encoded without change into 64 bit 'data' frames. The TCF sequence is transferred between the mobile facsimile machine and the facsimile machine in the PSTN and is conveyed across the link in data frames. In this way the quality of the end to end path including that of the radio environment is checked as there is no protection across the radio path for data frames other than the inherent transparent bearer service FEC and interleaving.

If for whatever reason the facsimile machines need to change their message phase speed, the fax adapters must 'look inside' the T30 DCS BCS frame to check the negotiated message speed and before the facsimile apparatus is allowed to change speed the air interface rate must be changed. This is known as channel mode modify.

In order to support the CCITT T30 message speed of 7200 bps, which is not a GSM rate, every 4th 64 bit data frame is packed with the 64 bit sync code.

To go into much more detail is beyond the scope of this text. For further reading see Reference 5.

12.7 Supplementary services

12.7.1 The point to point short message service (SMS)

The short message service allows alpha-numeric text messages to be sent to and from a mobile phone via a short message Service Centre (SC) using the SACCH (Slow Associated Control CHannel) rather than the TCH (Traffic CHannel). The SMS is therefore able to send and receive short messages irrespective of whether the TCH is being used for speech or data (or fax). If a speech or data call is not in progress then the short message is sent on the Stand Alone Dedicated Control CHannel (SADCH).

Routeing short messages around the PLMN is achieved through the use of an enhancement to the CCITT No 7 signalling system known as the Mobile

Application Part (MAP). MAP is used extensively for routeing of all traffic in the PLMN, not just short messages.

The size of a single short message is limited to 160 characters by virtue of the available capacity within MAP.

Short message data is protected on the air interface by the inherent control channel encoding and protocols. Those received by the mobile phone are normally stored on the subscriber identity module common in all GSM phones. At present SIMs can store up to five short messages which may be read or deleted at the user's discretion.

Outline of operation
Figure 12.13 shows the key functional network elements of the short message service.

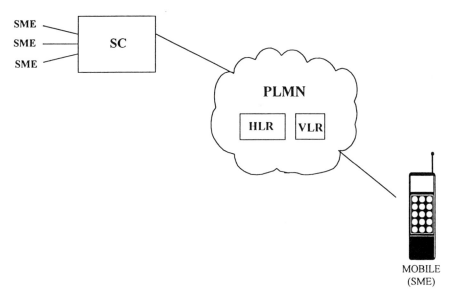

Figure 12.13 Short message service - key elements

Messages are sent via a Service Centre (SC). The point of origin and destination of a short message is known as a Short Message Entity (SME). An SME may be a GSM mobile phone or a device outside the PLMN such as a terminal on an X25 network.

The prime function of the SC is to receive short messages from an SME, store them and then attempt to deliver them to the receiving SME.

Once the SC has received the short message from the originating SME there is no requirement to maintain the communication link to that originating SME. The short message service is therefore essentially a store and forward service.

When an SC wishes to deliver a short message to a mobile, it must first interrogate the Home Location Register (HLR) to obtain routeing information for the short message upon receipt of which a delivery attempt may be made.

If the delivery of the short message is successful then the mobile will acknowledge its receipt.

It is possible for the mobile to be unavailable because it is perhaps switched off or is temporarily out of coverage. In such a case the delivery attempt will fail and the SC will retain the short message until the network informs the SC that the mobile is available. At the instant the delivery attempt fails, a flag is set for that mobile in the Visited Location Register (VLR) and a message waiting condition holding the SC address is set in the HLR. When the mobile becomes available, the VLR and HLR are aware of a pending short message and the appropriate SC is notified accordingly. The SC will then attempt another delivery. This feature is known as the short message 'alert'.

The GSM specification does not cover the design of the SC or the means by which it is connected to the PLMN. Neither does it specify how an SME in the fixed network communicates with the SC. It is the latter point which has recently received some attention by those wishing to promote the use of the short message service and work is currently in progress within GSM committees to examine standardisation of access protocols between the SC and users in the fixed network. At present access to SCs is typically by voice mail systems, X25 networks, packet radio networks and from the PSTN via modems. Figure 12.14 shows a realisation of the short message service being offered by Vodafone Ltd in the UK.

Short message data structure

Short messages are accompanied by a number of other parameters some of which are added by the network entities and some of which are accessible to the originator of the short message.

Figure 12.15 shows the format of four short message Protocol Data Units (PDUs) and their associated parameters. The Submit PDU is the format sent by the originating entity of a short message to send a short message.

The Deliver PDU is the format presented to the receiving entity of a short message.

The Command PDU is the format sent by the originating entity of a short message to request an action on a previously submitted short message

The Status Report PDU is the format presented to the originating entity of a short message indicating the status of a previously submitted short message (e.g. short message successfully delivered to its destination).

The Message Type Indicator merely indicates whether the PDU is a Submit, Deliver, Command or Status Report.

The Message Reference is a number assigned to a particular short message.

The User Data Header Indicator allows the user data to contain a header which can indicate that the user data field contains special information.

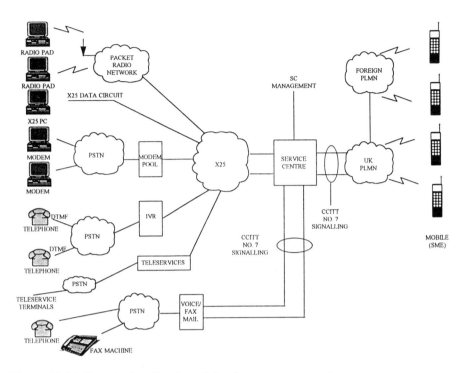

Figure 12.14 Practical realisation of the short message service

The Status Report Request indicates that a status report is required. The Status Report Indicator indicates that a status report has been requested.

The originating address and destination address formats allow a number of different types of address to be used, for example X121, E164, national, international, private etc.

In the case of sending a short message, it is possible for the originator to set the validity period of the protocol identifier which allows for some of the features described later.

The short message service caters for a number of alphabets and the alphabet being used is identified in the data coding scheme field . The default alphabet is a modified CCITT IA5 alphabet in which all the control characters except for carriage return and line feed have been replaced with European variants such as Greek characters, umlauts and accents. Only 7 bits are used per character and the eighth bit is not present at all. This allows 160 characters from the GSM default alphabet to be packed into the available 140 octets user data field. The data coding scheme can also indicate that the user data field contains binary or 8 bit data rather than the GSM default alphabet.

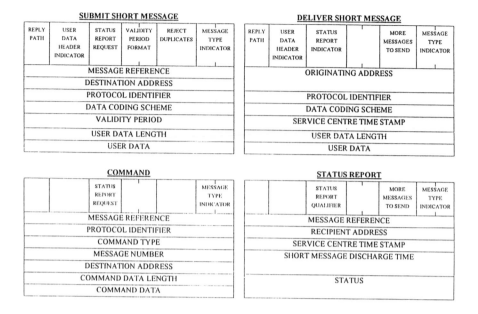

Figure 12.15 Short message protocol data unit types

When a short message is received by the mobile phone the short message service centre time/date stamp is also normally displayed together with the SC address and the originating address of the short message. The SC time stamp indicates the time at which the short message was received by the SC.

It is beyond the scope of this paper to describe in detail all of the parameters associated with a short message. A full description will be found in the GSM specification [4].

Mobile user considerations

Creating alpha-numeric text messages from the key pad of a mobile phone is impractical for some applications. Additionally, it is necessary for either the mobile or the user to insert some of the other parameters associated with a short message submit PDU as mentioned earlier. A user who merely wishes to send a simple text message from the mobile phone key pad is probably only concerned with the destination address, and possibly the service centre address. Default values for other parameters in the submit PDU are often set as default values in the mobile phone itself, allowing editing if desired.

For applications wishing to exploit more fully the capabilities of the short message service it is more practical to attach a terminal or PC to the mobile phone via the 'R' interface or via a PCMCIA card either of which overcomes the limitations of the key pad and display on the mobile phone. Unless the user has the need for a PCMCIA card for data or facsimile, a PCMCIA card is currently an expensive item often costing more than the mobile phone. A more attractive development is a proprietary cable or the 'R' interface.

GSM has defined a protocol for the attachment of an external terminal or PC to a mobile phone. The protocol is defined in Reference 7 and describes a 'block mode protocol' and a 'character mode protocol'. The block mode protocol makes provision for sending and receiving short messages, sending commands and receiving status reports and allows full access to all the short message service PDU parameters. A PC or some other intelligent device is required to implement the block mode protocol.

The character mode protocol merely allows the user to send the destination address and the short message text itself. All other parameters are those preset in the mobile phone or edited through the mobile phone keypad. A simple terminal emulator or dumb terminal is all that is required in this case.

Short message service features
A number of features are provided by the SMS specification (Reference 4). They are too many to cover in any detail here and so listed below are some of the most prominent.

- **Commands**: The mobile user may carry out a number of other operations on a short message which has previously been submitted (provided it is still held in the SC) by sending a Command PDU. For example a user may enquire on the status of a short message or delete a short message by using the message reference number and its destination address.

- **Replace short message**: Because of storage limitations for short messages in the mobile, the SC is able to indicate to the mobile that a short message is a replacement short message of a particular type for one which has been previously delivered to its destination address. In this way, messages conveying basically the same information may be updated without incurring additional memory usage in the mobile phone. The short message types are identified in the protocol identifier field. There are currently 7 types defined. Their use is at the discretion of the originator of the short message since the short message type is unique when associated with a particular originating address and SC address.

- **Display immediate**: The SC is able to indicate to a mobile that a short message is for 'immediate display' on receipt rather than for storage for displaying later at the user's request. Such a feature is useful where text information is to be conveyed between one user and another whilst a speech call is in progress between them. The indication is conveyed in the data coding scheme field through the use of the 'message class 0' parameter.

- **Ignore short message:** An SC may indicate to a mobile that the mobile may acknowledge receipt of a short message but may ignore its contents and not store it. Such a feature is useful when the alert feature described earlier is used merely to ascertain whether or not a mobile is available. The indication is conveyed in the protocol identifier field through the use of the 'short message type 0' parameter.

- **Memory capacity exceeded**: This feature allows a mobile (whose memory capacity for short messages was full and now has available space) to notify the SC so that any short messages awaiting delivery to that mobile may now be delivered.

- **Concatenated short messages**: This feature allows for a number of short messages to be sent where they form a message which is longer than the maximum in any one PDU, i.e. more than 140 octets. The user data contains a header which indicates that there is concatenation. The header also contains a sequence number and the total number of short messages which make up the whole message.

- **Data download:** This feature allows network operators or responsible administrations to down load data of a management nature to the mobile phone or the SIM. The indication is conveyed in the protocol identifier field.

- **Interworking with other services:**   A service centre may provide the capability to interwork with other services such as e-mail or facsimile. A user submitting a short message may mark the message for specific interworking by setting a particular protocol identifier value. For interworking with other networks such as X25, the user may indicate that the destination address of the short message is an X121 address.

12.7.2 The cell broadcast service

The cell broadcast service enables a GSM mobile phone to receive and display information which is broadcast on a regular basis. This service is similar to the UK television Teletext/ Ceefax services.

The information is first sent to a cell broadcast centre where it is administered and stored for onward transmission to a GSM base station from where it will be broadcast on a cyclic basis to mobile users. The repetition rate for broadcasting any particular topic from the base station is configurable by the network operator and may be between approximately 2 seconds and 32 minutes. As with the TV Teletext/Ceefax services, the refresh rate for updated information sent to the mobile is slow and gets worse the more topics there are to be broadcast. Information sent to the cell broadcast centre may be marked for broadcast over a general area or be geographic area specific. It is therefore possible to receive for example weather or road traffic information for the particular area in which the mobile user is travelling.

The mobile user will be able to select the information required through the phone keypad. There is no mobile user interaction with the base station or any other part of the PLMN for this service.

Each cell broadcast message can be up to 93 characters long although concatenation is possible for longer messages. The cell broadcast message has a number of parameters associated with it. One such parameter provides an indication that the information has changed since the last broadcast.

GSM has standardised the 'page' numbering for some of the more common topics such as weather reports and road traffic information.

The business case for freely broadcasting information of any value is difficult to justify. Revenue will undoubtedly come from the need for the mobile user to take some other fee paying course of action for more detailed information (e.g. a premium rate telephone call). Such a precedent already exists for some of the TV Teletext/Ceefax services.

12.8 Summary

With the advent of cellular mobile telephony it is now possible to extend office based speech and data communications to the mobile environment.

The problem with 'data' (excluding facsimile) is that there are too many standards to promote its widespread use in fixed networks. It is not surprising therefore that the use of 'data' in cellular mobile telephone environments is small. Standardisation of access for GSM mobiles should of course promote mobile applications although some mobile manufacturers have already chosen to deviate from the ETSI standards for what appear to be valid commercial reasons. Such deviations however if perpetuated risk plunging GSM into the same confusing world as fixed networks.

The use of 'data' in mobile environments is in the main inextricably linked to the availability of fixed network services. The current popularity of Internet seems to be focusing mobile users towards conventional mobile data using products and software available to support applications such as file transfer and electronic mail over TCP/IP. There is currently no special provision made in GSM standards for TCP/IP.

Other than speech, facsimile is currently the most commonly used method of communications world-wide. The reason for the success of facsimile is the ease of use, due primarily to its simplicity, standardisation as an end to end service and the absence of any alternative means of information transfer. Delay in widespread implementation of data networks such as ISDN has merely fuelled the development of sophisticated higher speed modem technology for facsimile machines and the

PSTN in general and has raised serious questions concerning the future of group 4 facsimile. The demand for mobile facsimile is as yet uncertain.

For data applications such as text messaging the short message service is providing for an attractive alternative conventional data where real time data transfer is not a requirement.

The short message service has the advantage that it offers a total solution which is both simple to use and cost effective. It also provides virtually guaranteed confirmed delivery of messages to their final destination.

For those applications for which real time data transfer is a requirement and where a circuit switched call is unsuitable then the GSM packet radio service may provide the answer.

References and further reading

1   IS4335 International Standards Organisation

2   Bleazard, G.B. *NCC Handbook of data communications.* NCC publications

3   Jennings, F. *Practical Data Communications Modems, Networks and Protocols* (Blackwell Scientific Publications)

     CCITT Recommendations V24, V28, V25bis, X25, X28, T4, T30

     GSM Specifications as follows:-
     03.10 GSM PLMN connection types
     03.38 SMS Alphabets
     03.39 SMS Service Centre access protocols from the fixed network

4   03.40 Short Message service
     03.41 Cell Broadcast service

5   03.45 Transparent Facsimile Service

6   03.46 Non Transparent Facsimile Service
     04.21 Rate adaptation at the MS BSS interface
     04.22 RLP
     07.01 General TA functions for mobile stations
     07.02 TA Functions for the asynchronous services
     07.03 TA functions for synchronous services

7   07.05 SMS access protocols at the mobile station
     07.06 Setting mobile station parameters
     09.xx PLMN/fixed network Interworking

Chapter 13

Future satellite
mobile telephone networks

Bruce R Elbert

Introduction

In 1990, a paradigm shift occurred in the satellite communications industry when Motorola first introduced their low-earth orbit satellite (LEO) phone system called Iridium. Up to this time, the shared belief was that the mainstay geostationary earth orbit (GEO) satellites exemplified by those of Intelsat and Inmarsat, would not support direct links to hand-held phones. Then in 1994, several companies devised schemes to provide mobile service to hand-held phones, employing literally all altitudes ranging from Iridium's 800 km all the way up to 36,000 km at GEO. This chapter identifies the particular technical problems and challenges which the successful satellite mobile service must address. Among these are the selection of an appropriate orbit and constellation, the trade-off among modulation and multiple access methods, and frequency reuse strategies to permit a cost-effective service without sacrificing bandwidth. This chapter also considers which of the planned systems has the greatest likelihood of implementation and commercial success.

13.1 Satellite communication fundamentals

Satellite communication networks have been in existence for nearly three decades. Once thought to be an advanced and exotic technology, communication satellites have been reduced to the mundane. To understand how pervasive satellite links have become, drive through any neighbourhood and count the Sky TV dishes. In the field of mobile communications, however, satellites are gaining ground because of their unique ability to extend service to places that the terrestrial infrastructure currently cannot. Satellites already play a key role in maritime communication and are about to take on the vehicular and hand-held radiotelephone markets as well.

13.1.1   Radio frequency spectrum availability

Radio spectrum has been allocated by the International Telecommunication Union (ITU) to satellite communication, just as it has to fixed radio services like microwave, mobile radio, and broadcasting.   Figure 13.1 illustrates the usable part of the radio spectrum, ranging from 1 MHz to 100 GHz. Frequencies from 1 GHz to 30 GHz have already been exploited for satellite communication. It is common to use the old letter designations in reference to the satellite bands; thus L-band (~1.5 GHz) is the segment of greatest interest for mobile services.   Cellular networks were first established below 1 GHz, but now we see them 'encroaching' on satellites as the demand for spectrum escalates.   Likewise, Mobile Satellite Service (MSS) needs are being satisfied above the L-band in the segment known as S-band (~2.5 GHz). The higher frequencies in the C-, X-, Ku- and Ka-bands are not intended for mobile applications.

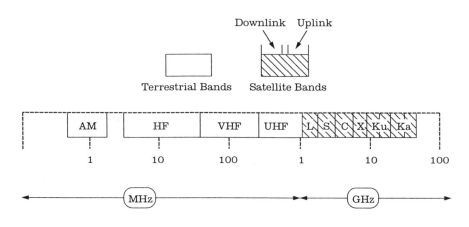

Figure 13.1   Radio spectrum

An additional complication, as indicated in Figure 13.1, is that much of the satellite spectrum must be shared with terrestrial services. Sharing is a complex issue; terrestrial services with simple inexpensive antennas can easily interfere with satellite services and be interfered with by them. However, the growing demand for spectrum means that sharing will continue to be a way of life.

13.1.2   Progression of commercial satellite development

Over the years, satellite designs have evolved to meet the demand for more effective services at lower prices.   Launched in 1965, the first commercial communications satellite, Intelsat I, was very small (Figure 13.2). It could only relay a single TV channel or a few hundred telephone channels. Its first application was to establish a link between Andover, USA and Goonhilly Downs, UK. Both stations required antennas of approximately 30 meters in diameter to capture enough signal from the relatively weak satellite. Since that time, satellites have grown in power, allowing many more stations to connect over the network and increasing capacity from hundreds to thousands of telephone channels.   In the 1990s, satellites deliver comparable services to very small aperture terminals (VSATs), which have a diameter of 1.2 meters and cost less than £8,000 each. The bottom line is that satellites have grown in physical size, power, complexity and cost to reduce the size and cost of the earth station.   The penetration of satellite services into businesses and homes has increased as a direct result.

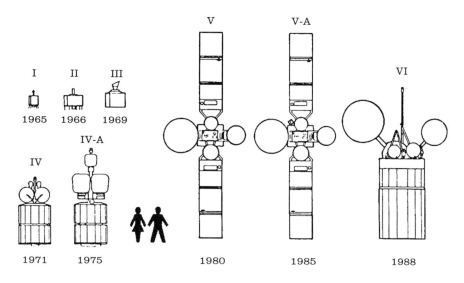

Figure 13.2   A progression of communications satellites (the INTELSAT series from I to VI)

13.1.3   Range of satellite services

The frequency allocations made by the ITU divide satellite services into three broad categories: Fixed Satellite Service (FSS), Broadcasting Satellite Service (BSS) and Mobile Satellite Service (MSS). This is done with the aim of creating a homogeneous sharing of the bands and thereby increasing the overall usage. Applications generally follow these divisions.   However, occasionally some enterprising individual or organization finds a way to slip an application from one category into the more-available or less expensive part of the spectrum.

Fixed satellite service
The FSS is really the 'grandad' of all the satellite services. With the low power of early satellites, earth stations had to be large and fixed on the ground.   The C-band spectrum in the 4 to 6 GHz range is good for line-of-sight propagation, but not very favourable for mobile service.   The Ku-band was developed in Europe for the

FSS and followed in the US as a means to expand capacity. The FSS succeeded in establishing the utility and economy of satellite service and is still heavily used for domestic and international communications.

Broadcasting
The early thinkers in satellite systems, not the least of whom was Arthur C. Clarke, envisioned that satellites could broadcast radio and TV programmes directly to the public. Europeans were prominent in promoting this idea at the first World Administrative Radio Conference (WARC) on BSS, held in Geneva in 1977. At that conference, every country in the world was assigned at least one orbit position for broadcasting satellites in the Ku-band region. While it has taken a long time for the technology and business to catch up, BSS systems offer an alternative to over-the-air stations, cable TV, and cassette rental.

Mobile
Comsat and Inmarsat chose to use the L-band spectrum allocated to MSS. It proved to have the right propagation characteristics and the service is so successful that all land radio stations (e.g. the Morse code operators) have been shut down in favour of Inmarsat. From this foundation, a transition is occurring to land-based MSS, or LMSS, as satellite power increases and the cost of mobile terminals decreases. For example, Peter Arnett of CNN transmitted his Gulf War reports from a position inside Baghdad using an Inmarsat Standard A satellite phone that he smuggled into the country. He used a collapsible umbrella-style antenna to close the link. Now, Inmarsat has introduced the Standard M phone, which uses digital voice compression and is available in an attaché-case sized unit.

In reality, MSS is only beginning to sprout. Motorola surprised everyone when it announced Iridium, the first proposal for a global satellite network to serve hand-held phones. This caused a real paradigm shift in the satellite industry. Now, several companies have proposed doing the same, including Inmarsat. The issues that confront the developers of such networks are discussed later in this chapter.

13.1.4 Satellite links

The link between earth station and satellite is like other microwave links, exhibiting a variety of propagation modes and degradations. As with cellular radio, adding mobility tends to make the situation more complex.

Figure 13.3 illustrates the familiar radio link from transmitter to receiver. Below the diagram is the power balance equation, expressed in dB, to allow the substitution of addition and subtraction for multiplication and division. The satellite industry practices 'dB artistry' in the same manner as those who are already involved with terrestrial radio systems. The satellite is nothing more than a very distant microwave repeater station or relay. The power balance simply states that the received power equals the transmitted power, plus all gains and minus all losses. In FSS and BSS service where the earth station does not move, the link is very stable, affected progressively by rain attenuation at frequencies above 4 GHz, and some ionospheric and atmospheric phenomena. The result is a relatively stable link dominated by Gaussian noise. However, a link margin in the range of 1 to 6 dB assures reliable service. The type of analysis used to evaluate these factors is called a 'link budget'.

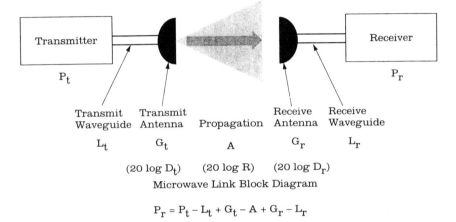

Microwave Link Block Diagram

$$P_r = P_t - L_t + G_t - A + G_r - L_r$$

Power Balance Equation

Figure 13.3   The radio link power balance (expressed in dB)

Consider the specific arrangement in Figure 13.4 of the single-hop MSS link between a mobile subscriber using a hand-held phone and a fixed gateway hub station that connects to the PSTN. The mobile user transmits and receives within the L-band uplinking to the satellite in the 1.6 GHz band segment and downlinking in the 1.5 GHz segment. At the satellite, the relatively weak signal is received by a large diameter antenna, amplified and translated to the feeder link band, which is assumed to be at Ku-band. The satellite transmits the signal at 11 GHz; the gateway transmits at 13 GHz. Specialised base station equipment recovers the channel and a mobile telephone switch provides access to the PSTN. The complete hop from the gateway to the mobile subscriber is called the 'forward link' and the reverse is called the 'return link'.

Coming back to one particular aspect of the satellite link, indicated by the equations below the block diagram in Figure 13.3, there are three factors related to the square of the frequency: the gain of the transmitting antenna, the path loss, and the gain of the receiving antenna. If the size of the transmitting antenna is kept constant and the operating frequency is increased, the received signal at the satellite also remains constant. This is because the increase in gain is exactly compensated by the increase in path loss. The same situation applies to the path from the satellite to the receiving antenna on the ground. The path loss, *A*, is also proportional to the square of the distance (20 log R), which would appear to put satellites in the geostationary earth orbit (GEO) at a disadvantage relative to satellites in low earth orbit. However, it is possible to compensate for this by increasing transmit power and/or antenna gain.

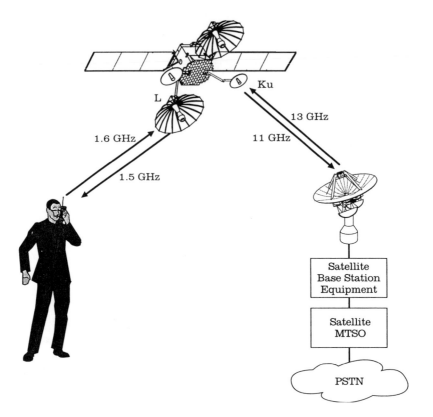

Figure 13.4   The mobile-to-fixed link

Figure 13.5 shows a typical link budget for the forward and return links. This budget assumes the use of voice compression capable of reducing the rate to 4.8 kbps. In the forward direction, the gateway earth station transmits the channel at 13 GHz with a 9 m antenna. Combining the transmit power and gain of this antenna, achieves an Effective Isotropic Radiated Power (EIRP) of 46 dBW. The path loss in the uplink to the satellite is 206.7 dB. The satellite receive gain, minus the system noise temperature, termed the G/T, is -3.6 dB/K, producing a value of uplink C/N within the noise bandwidth of one channel of 28.9 dB. This must be combined with assumed interference from other sources (to be discussed later) which collectively produce a C/I of 23.8 dB. The resulting C/N for the uplink is

23.3 dB. This is only half of the problem because this channel is amplified, translated in frequency, and radiated to the gateway earth station.

4.8 Kbps Voice	Forward (FES to MET)		Return (MET to FES)	
	Up	Down	Up	Down
Frequency, GHz	13.0	1.55	1.65	10.8
Antenna	9.0 m	–	(Mast)	–
EIRP, dBW	46.0	28.0	12.5	4.5
Path Loss, dB	–206.7	–188.2	–188.7	–205.1
G/T, dB/C	–3.6	–16.0	–0.3	34.5
C/Nth, dB	28.9	17.0	16.7	27.1
C/IM, dB	30	22	50	24
C/I	23.8	19.6	23.4	21.5
C/N link, dB	23.3	15.2	16.0	21.1
C/N total, dB	–	14.6	–	14.8
C/N req, dB	–	11.9	–	11.9
Margin, dB	–	2.7	–	2.9

Figure 13.5   Typical L-band link budget - vehicular telephone

Examining the downlink, the operating frequency at L-band is 1.5 GHz. Again, consider the combination of the satellite antenna gain and the transmit power, which produces an EIRP of 28 dBW in the direction of the gateway. Path loss at L-band is less than at the Ku-band frequency by the term 20 log(13/1.5), or 18.5 dB. The mobile phone in this example has a G/T of -16 dB, which is very poor in comparison to even the smallest of dish antennas used in the BSS. The main reason for this is that the gain (G) is typically less than 10 dB to provide the small size and broad beam associated with mobile radio antennas. In fact, -16 dB is high compared to what the hand-held phones produce. The L-band downlink produces a C/N of 17 dB, which when combined with the C/I, results in a combined C/N of 14.6 dB. In this particular system, the transceiver of the mobile phone is looking for a C/N of 11.9 dB for reliable service, allowing for a link margin of 2.7 dB.

The link budget for the return direction is featured to the right of Figure 13.5. Following exactly the same progression, the link provides a margin of 2.9 dB, very comparable to that of the forward link. The significance of this margin is that this link will work satisfactorily for line-of-sight conditions. Any significant shadowing or antenna misalignment will push the link below threshold. One of the unfortunate properties of satellite links is that they cannot produce the levels of margin that are customary in cellular networks. To talk about margins of more than 16 dB is to push the technology and investment cost essentially to the limit. The result is that users of satellite phones can probably be expected to 'co-operate' more to find those locations and situations that prove satisfactory. For example, a user would find a position on the ground with a clear path to the satellite. Once established, the link will work reliably for as long as the user wishes to speak.

13.1.5   Frequency co-ordination

The design of a modern MSS satellite provides spectrum reuse by placing an array of multiple spot beams over the service area.   Each spot is analogueous to a terrestrial cell, except that it is much larger in geographic terms.   The other basis for sharing is the longitude separation of satellites that use the same frequency band.   This is practical for the traditional FSS and BSS systems, as illustrated at

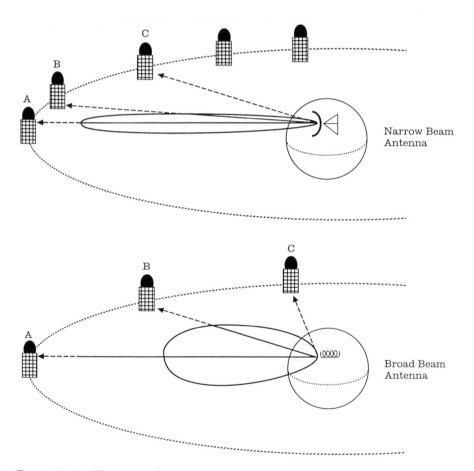

Figure 13.6   How ground antennas share spectrum

the top of Figure 13.6.   In this example, the earth station dish has its main beam directed toward satellite A;   satellite B is roughly at the 3 dB point on the beam, while satellite C is located at the start of the sidelobe region and therefore is not subject to interference.   The rest of the satellites are likewise safe.

The situation at L-band for mobile services is less attractive from this standpoint, as illustrated at the bottom of Figure 13.6.   The earth station antenna in this case is nothing more than a helix, chosen for a broad beamwidth. The reason for doing this is that the broad beam is not sensitive to the orientation of the antenna, either in azimuth or elevation. The closest orbit position that can reuse the spectrum is all the way over to satellite C, which is effectively over the horizon from the user. To this class of user, MSS offers no frequency reuse provided by orbit separation. The only reuse possible is due to the manner in which the satellite places its coverage area over the earth. Two satellites that have non-overlapping coverage beams will not interfere even if they are at the same orbit position.

The specific spectrum that is available at L-band is shown in Figure 13.7. The lightly shaded downlink and uplink bands are already in use by Inmarsat and others, while the cross hatched bands are reservations for the coming generation of non-GEO satellites that could be launched around the end of this decade. Notice from the footnotes to the table that the allocation has quite a few stipulations about usage. Some portions give MSS a primary status, meaning that the non-primary services, called secondary services, must give way if interference is possible. On the other hand, MSS is not even listed as secondary in the aeronautical bands, indicated by the letter A. Current practice is for MSS systems to use these bands until aeronautical mobile satellites are launched.

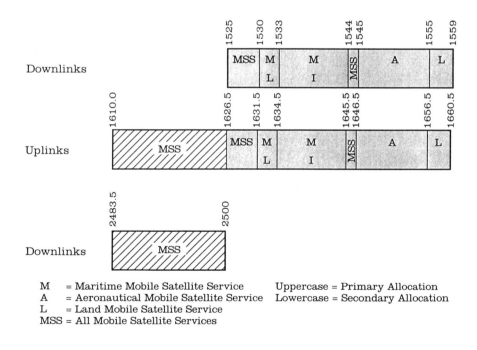

M	= Maritime Mobile Satellite Service	Uppercase = Primary Allocation
A	= Aeronautical Mobile Satellite Service	Lowercase = Secondary Allocation
L	= Land Mobile Satellite Service	
MSS	= All Mobile Satellite Services	

Figure 13.7   Mobile satellite services spectrum

13.2 Orbits and constellations

A number of new entries into the MSS marketplace are developing the concept of a global network, based largely on the use of non-GEO orbits.   Figure 13.8 illustrates three orbit configurations that could provide global service.   Each orbit brings with it operational and financial considerations.   The polar regions are excluded from consideration as these markets are extremely small.

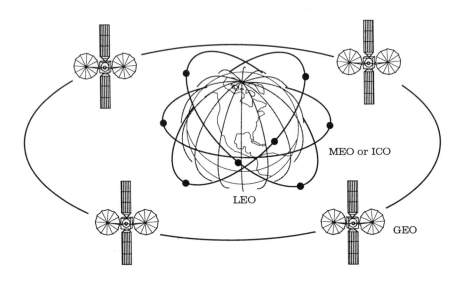

Figure 13.8   MSS orbits

13.2.1   Geostationary earth orbit (GEO)

All the extensive experience to date with commercial satellite communication has been at GEO. The overall simplicity of the arrangement and the ability to use fixed antennas on the ground have allowed GEO to reach critical mass in some key industries. However, the antennas used by mobile telephones are non-directional, so the fixed nature of the GEO satellite has less importance. In addition, there is the consideration of the longer path length and greater path loss, coupled with the increased propagation delay, as compared to MEO and LEO. The net result is that the GEO approach is attractive because it requires only one satellite to start and three to serve the world, while there is a debate as to the impact of increased delay.

13.2.2 Non-geostationary earth orbit

The non-GEO orbits, shown in Figure 13.8, give up the synchronisation of satellite revolution for the advantages of being closer to the earth. These advantages include reduced propagation delay and path loss. In exchange, one needs to deploy significantly more satellites to maintain continuous coverage of the user for real-time applications like telephony. An alternative would be to sacrifice real time service by providing short periods of data transfer or store and forward messaging. The economic viability of the latter has never been demonstrated. However, no one can dispute the attractiveness of addressing the mobile telephony market, provided that service quality and cost meet the expectations of subscribers.

13.2.3 Lower earth orbit (LEO)

The LEO approach generally locates the satellites in closest proximity to earth, typically 700 to 1000 km, with an orbit period of about 90 min. This provides the minimum time delay and propagation loss, but increases the number of satellites required, for continuous service, by a factor of ten. Hand-off of calls is also a consideration. This type of system would be expensive to implement and complex in its management. There is also a major concern about the ability of the system to provide continuous service due to the effects of terrain blockage.

13.2.4 Medium earth orbit (MEO)

The requirement to have many satellites, with their attendant cost, has produced some alternative system concepts, where the orbit is raised to the 7,000 to 15,000 km range. Anything less would subject the satellites to the harsh environment of the Van Allen belt. Many fewer satellites (16 or less) are needed and the delay is less than half that for GEO. Path loss is also 6 dB less than GEO because of the square law nature of propagation. The continuity of service is better than LEO because of the longer duration that a given satellite is in view of the user and gateway earth station. The intermediate circular orbit (ICO) is a specific arrangement of MEO, recently adopted by Inmarsat for the planned P system. The satellites do not pass over the poles, but are inclined, to provide the best service to populated regions. An environmental concern has been raised about the eventual re-entry of satellites through the earth's atmosphere.

13.3 Mobile satellite systems

This section supplies details on Mobile Satellite Systems (MSS) currently in existence and those planned. An indication of the types of terminals that can be employed with these systems is provided in Figure 13.9.

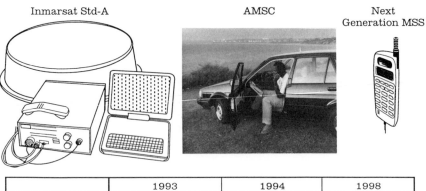

	1993	1994	1998
Terminal Size	Transportable	Mobile/Vehicular	Hand Held
Terminal Cost	$15,000 to $25,000	$2000	$1000 to $1500
Usage Cost, $/min	$3.50 to $6.00	$0.90 to $1.50	$0.50 to $3.00

Figure 13.9   Evolution of land mobile satellite phones

13.3.1   Existing systems (GEO)

The following GEO systems were in existence at the time of this writing.   They represent the most real category of MSS programmes.

Inmarsat

The Inmarsat system evolved from an entrepreneurial start by Comsat in the late 1970s. Comsat used its win of a US government contract to launch into service the first maritime mobile satellite system.   The first satellites, called Marisats, where built by Hughes and are still in operation today.

Inmarsat was organised by Comsat as an international joint venture of governments and telecommunication operators, much like Intelsat.   Figure 13.10 provides a summary of Inmarsat characteristics.   A progression of Inmarsat terminal types is provided in Figure 13.11, beginning in 1982 with the classic Standard A analogue terminal for ship-board use.   The antenna in this case is a parabolic dish about one metre in diameter, with a gimbal system to maintain pointing during ocean travel.   Inmarsat M is the first example of a true land-mobile telephone service.   By 1996, Inmarsat will introduce the mini-M terminal, which promises to be the size of the first generation of portable cellular phones.

The performance of Inmarsat GEO satellites has grown in terms of power, culminating with the Inmarsat III series now being deployed.   The spot beams shown in Figure 13.12 will allow portable terminals to access the satellite directly. This will be similar to the first portable cellular phones, which were about the size of a loaf of bread. Inmarsat P, on the other hand, will be directed toward the hand-held market of the next decade.

- International consortium created to provide maritime mobile communications

- Inmarsat II provides global beam coverage

- Inmarsat III to provide both global beam coverage and regional spot beams over land masses

- Inmarsat P-21 study examines feasibility of providing global hand held personal communication

---

Figure 13.10   Inmarsat - the established service provider

1982	Inmarsat-A	Original voice/data terminal
1990	Inmarsat-aero	Aero voice and data
1991	Inmarsat-C	Briefcase data
1993	Inmarsat-M	Briefcase digital phone
1993	Inmarsat-B	Digital full service terminal
1994	Global paging	Pocket sized pagers
1995	Navigational services	Variety of specialized services
1998	Inmarsat-P	Hand held satellite phone

Figure 13.11   Inmarsat standards and services

Inmarsat is pursuing a global MEO strategy with a new affiliated company called I-CO Global Communications, Ltd. They will offer hand-held and mobile telephone service in competition with the global systems described later in this chapter. The investors in I-CO include many of the Inmarsat signatories.

AMSC and Telesat Mobile

American Mobile Satellite Corporation (AMSC), reviewed in Figure 13.13, is a public corporation founded by the first applicants for domestic LMSS in the US. AMSC began service in 1995 with the launch of a dedicated GEO satellite and the start of operation of the associated ground infrastructure. The latter interfaces with the PSTN and interoperates with domestic cellular systems. The main market for service, determined by the technical capability of the network, is vehicular telephone service in areas not covered by cellular. The phones are dual mode, thus being able to complete calls over the cellular network when in range of a base station. Otherwise, the AMSC mobile telephone will seek the satellite.

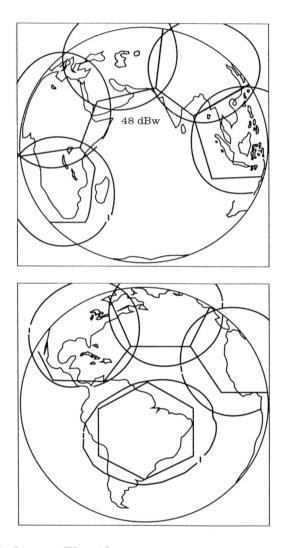

Figure 13.12　Inmarsat III spot beam coverage

AMSC initiated operations a few years previously with an Inmarsat C hub for position location and low speed data service for trucking companies. Interestingly, their satellite was an old Marisat that put Inmarsat into business in 1982.

As a public company, AMSC is responsible for returning a profit to its founders and to the public stock holders. The stock is currently traded on NASDAQ as SKY-C.

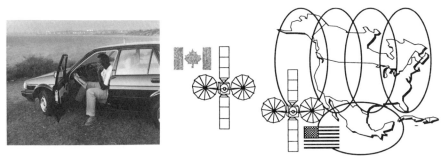

- Currently licensed as provider of mobile satellite services in U.S.
- Shared technology with Canada to enable mobile communications across North America
- Voice, data, vehicle location, rural telephone, emergency communication
- Major equity partners: GM Hughes Electronics, McCaw Cellular, Singapore Telecom, and Mtel Corp
- 2 GEO satellites: 1 U.S., 1 Canadian
- Interim service operating on leased capacity; first satellite launch in 1995

Figure 13.13   American Mobile Satellite Corporation (AMSC)

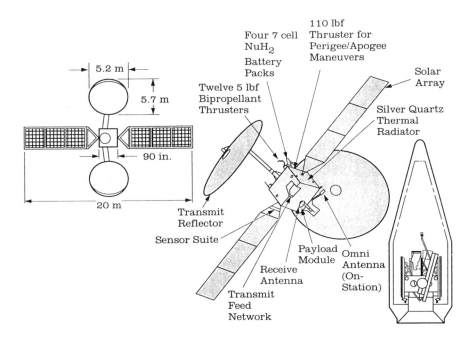

Figure 13.14   AMSC satellite

The satellite employed by AMSC and by the Canadian MSS operator, Telesat Mobile, is shown in Figure 13.14.   This is a high power design with dual L-band antennas, one for transmit and one for receive, each approximately 5 metres in diameter.   A small Ku-band antenna provides feeder links to the gateway earth stations.   The satellite was designed and built by a team consisting of Hughes Space and Communications Company of the US and Spar Aerospace of Canada. Figure 13.15 shows the antenna coverage pattern from the AMSC satellite, which includes the four continental US beams and other beams pointed at Alaska, Mexico and the Caribbean, and Hawaii.   The eastern and western US beams share the same frequency band segment, which is possible because of the wide separation.   The two central beams are in different band segments to avoid RF interference. Figure 13.16 shows a simplified block diagram of the repeater.   At the top is the forward direction while the return is shown at the bottom.   A direct Ku to Ku connection permits network control and supervisory information to flow between gateway earth stations.

The first generation AMSC satellite addresses a market composed mainly of automobiles, lorries, coastal vessels, and other vehicles.   Customers will include the general public who obtain the service through their cellular telephone companies, as well as government users.   In the next generation, AMSC will increase the performance of the satellite to allow service to hand-held terminals. The timing of this has not been announced.

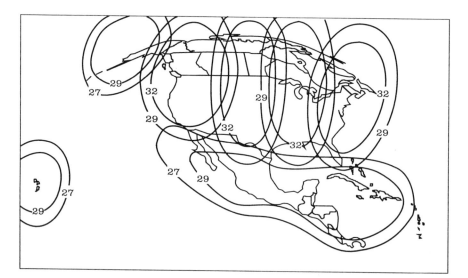

Figure 13.15   Typical AMSC L-band antenna gain, dBi

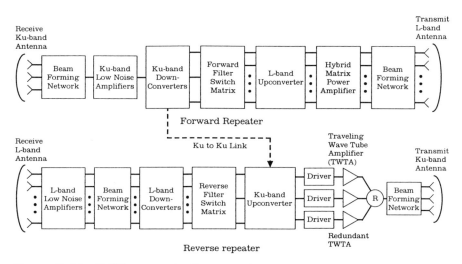

Figure 13.16   AMSC communications repeater

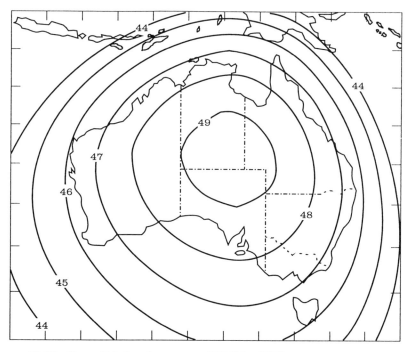

Figure 13.17   Optus B L-band coverage (EIRP in dBW)

Optus Mobile Sat

The Australian satellite communications provider, Optus Communications Ltd., began providing service to mobile customers in 1994. Optus B is the name given to the second generation of Australian satellites that includes an L-band repeater. With a relatively small antenna system and low amplifier power, Optus B is limited in capacity and performance. However, it has the ability to service up to 500 simultaneous subscribers per satellite. Figure 13.17 illustrates the single beam coverage of Optus B. From anywhere in Australia, users with vehicular terminals can obtain mobile telephone service. In such a large and lightly populated country, Optus B MSS service is gaining in popularity because terrestrial cellular is restricted to the larger metropolitan areas and roadway corridors.

Figure 13.18 illustrates an important consideration in the design of MSS networks. This graphic plots contours of constant elevation angle from the ground. For example, in the area around Sydney, the elevation angle in the direction of the satellite is approximately 55°. This is relatively high, meaning that the takeoff angle from the mobile antenna will clear most obstacles. Low elevation angles can subject the user to blockage from local terrain. A particular benefit of the GEO approach is that this elevation does not vary for a fixed location on the ground.

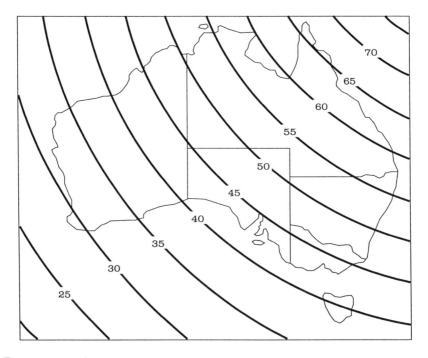

Figure 13.18   Optus B ground elevation angles

13.3.2 Systems under development (non-GEO)

Several non-GEO satellite networks for MSS have been proposed and some are under rapid development. The leading contenders among them are Motorola's Iridium, Loral's Globalstar, and TRW's Odyssey; Inmarsat's I-CO company was covered in the previous section. Only Iridium is proposed by a true cellular company, while the others come from traditional satellite communications companies. As of this writing, Motorola is considerably ahead of the others in organization, financing, technology development, and marketing. As a cautionary note, it is extremely difficult to determine the actual status of expenditure and roll-out of any system as there is usually a tight seal on this kind of information.

Figure 13.19 shows the basic orbital arrangement of Iridium and Figure 13.20

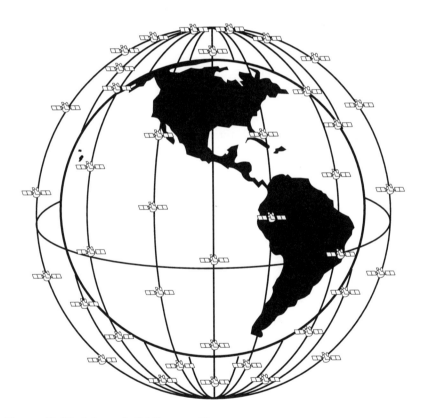

Figure 13.19　Motorola Iridium satellite system

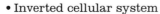

- Inverted cellular system
- 66 polar orbiting satellites; 11 satellites in 6 planes
- Phased array antenna
- Onboard processor
- 48 beams
- Intersatellite links
- Ka-band feeder links
- TDMA
- Service to handheld phones

Figure 13.20   Iridium system characteristics

presents some of the basic characteristics of the system. The plan is to have six polar orbital planes (60° apart), with 11 satellites in each. This architecture affords 100% coverage of the globe, including the poles. In fact, the polar coverage is better than that over any other region. Satellites are constantly in motion relative to the earth and interconnected through a system of intersatellite links. This substantially increases the complexity of the network. Furthermore, each satellite is to have 48 spot beams generated by a phased array antenna. The links to the mobile terminals are at L-band and are bi-directional in nature, using the new allocation identified in Figure 13.7. Feeder and intersatellite links operate at Ka-band. Importantly, Iridium represents the first proposal to offer ubiquitous service from space to hand-held phones.

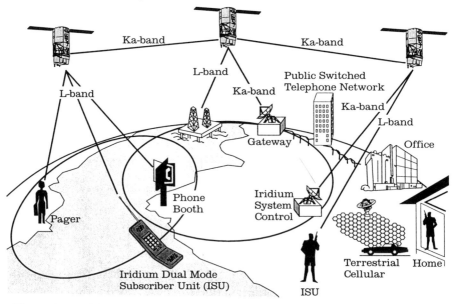

Figure 13.21 The Iridium network

The Iridium network, illustrated in Figure 13.21, will use these satellites to connect directly to users anywhere in the world. This is another unique feature: a hand-held phone in Africa can communicate directly with a hand-held phone in Asia (or anywhere else) without passing through ground facilities. Alternatively, the network supports calls to the PSTN and terrestrial cellular subscribers.

Motorola is the prime contractor for the entire system, but has spun off the business to a company called, not surprisingly, Iridium, Inc. Their approach to creating the business is to sell shares to companies and governments around the world. Later in this chapter there is a discussion of how this type of network stacks up against the GEO approach, from a financial point-of-view.

The Globalstar system is also in the developmental phase and represents a potential competitor to Iridium and I-CO. In December 1994, Global floated an initial public offering that garnered enough money to begin construction. They have built a strong team of partner companies around the world, including France Telecom, Airtouch and Alcatel, among others, and are in the process of gaining access to key domestic markets. Thus far, the US Federal Communications Commission has granted a licence to Globalstar, Iridium and Odyssey to operate their systems. While Odyssey is considerably behind the other two, it recently received a US patent on the use of MEO for mobile telephone service.

13.4 Multiple access and modulation

The selection of the multiple access method is always an important technical and operational decision, because it can influence critical parameters like service

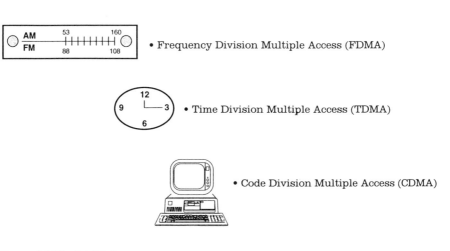

Figure 13.22 Multiple access systems

availability, system capacity, and network connectivity.   From a financial standpoint, capacity is very critical since it directly determines the number of users that a fixed investment space can support. As shown in Figure 13.22, the standard choices are frequency division multiple acces˜ (FDMA), time division multiple access (TDMA), and code division multiple access (CDMA). FDMA signals are arrayed on individual frequencies (similar to a radio dial), and appear and disappear as users employ their respective terminal devices.   In TDMA, there are fewer transmissions (carriers) visible, but each is the result of several users time-sharing the same channel i.e. two users cannot be on the air at the same time. Finally, in CDMA, users transmit on the same channel, at the same time, but with different digital codes that isolate the signals from each other.   There are two versions of CDMA: a narrow band case and a wideband case, the latter corresponding to each user transmitting across the entire piece of assigned spectrum.   Radio communication systems have used each of these for at least 30 years, so there is really nothing new here. Rather, it is how effectively the scheme performs under specific conditions (and these vary greatly) and how the associated hardware can be implemented within, say, portable and hand-held phones.

FDMA

The first multiple access scheme to be applied to satellite communication and the one with the most established track record (in analogue cellular, as well) is FDMA. There are many reasons why FDMA continues to have a strong following.   Figure 13.23 lists three of the key benefits. Foremost, each user transmission is separate from every other by using a different frequency. A user terminal can receive a selected signal by simply tuning to the associated frequency.   It is even possible to implement different types of transmission within the same system as long as they operate on their own channels. The frequencies are assigned to users from a pool that is under the control of a central management resource, also in direct contact

6 to 30 kHz

• Flexible
• Widely used in satellite and terrestrial networks
• Low complexity

Figure 13.23 FDMA

with all users through a common signalling channel.   This principle has been applied to satellite telephone networks for the past 25 years and makes good use of the common relay point afforded by the satellite itself.

An MSS FDMA network and an FDMA terrestrial cellular counterpart have similar properties when it comes to multiplying the available channel space or bandwidth. Cellular networks accomplish this by reusing frequency channels across geography by means of a cell plan. The available frequency band (in the range of 10 to 30 MHz) is typically divided among seven cells so that adjacent cells do not use the same frequencies. An MSS FDMA network uses the same reuse technique, but the cells are created by the beams from a multi-beam satellite antenna. With at least two full beams of separation, a pair of beams can reuse the same frequencies just like the geographical separation of cells in the terrestrial counterpart. This kind of frequency reuse arrangement effectively overcomes one of the most serious problems with FDMA, namely that you cannot allow two channels to lie on top on one another. In CDMA, this rule is overridden at the expense of appearing to waste bandwidth.

FDMA can be practised with analogue or digital modulation because the individual carriers are completely separate from one another. The most popular analogue modulation is frequency modulation (FM) because it can operate in a narrow bandwidth in the presence of various kinds of noise and interference. FM is the most prevalent modulation method in use for terrestrial cellular systems (e.g. AMPS in North America, and TACS and NMT in Europe) and it continues to have a following on satellites as well. There is a trend toward digital processing and modulation to reduce the bandwidth occupancy per telephone user, as will be discussed later. In terms of hub utilization, FDMA requires a separate channel receiver per active mobile user. Each frequency and user requires one channel unit at the hub. Users share other parts of the hub including the antenna, transmitter, switching equipment, and support systems.

TDMA

TDMA was first demonstrated on a satellite link in the late 1960s and has some

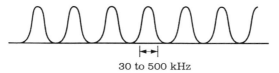

30 to 500 kHz

- Inherently digital
- Less hub equipment
- Supports frequency hopping

Figure 13.24 TDMA

important benefits for MSS (summarised in Figure 13.24). The premise behind TDMA is that multiple users share a somewhat wider frequency channel and the associated hub equipment. To accomplish this in a non-interfering way, mobile terminals transmit their information in the form of bursts of high-speed data at non-overlapping times. The total channel speed, and therefore bandwidth, is roughly proportional to the number of users sharing the same channel. TDMA is currently in use in very small aperture terminal (VSAT) satellite networks and is the core of the GSM digital cellular standard. Likewise, the currently operating North American digital cellular standard, IS-54, employs TDMA. Motorola has further advanced TDMA by adopting it for their Iridium project.

TDMA is inherently digital because information is stored and subsequently transmitted at high speed in the form of a burst. The technique demands precise timing among users that share the same channel. The necessary synchronisation is accomplished with circuitry at the hub; it must lock up to individual transmissions from the mobiles. In reality, there are several TDMA channels in operation at the same time, each on a different frequency. Combining all users on the same channel increases the speed and bandwidth, which pushes up the power demand from mobile terminals. Also, wider bandwidths in TDMA are more susceptible to the form of multipath fading that is frequency dependent i.e. selective fading. Multiple carriers therefore interpose a form of FDMA on top of TDMA. A benefit is that users can 'hop' between channels to access a particular hub, or to avoid interference. Hopping also facilitates seamless hand-off in non-GEO networks.

CDMA

CDMA approaches multiple access in a unique way by transmitting signals literally one on top of another. Such direct jamming would render FDMA and TDMA useless, but for CDMA this kind of interference is controlled because signals make use of orthogonal codes that are, by nature, uncorrelated. The code is in the form of a random bit sequence that is multiplied by the digitized data. The data rate of this random bit stream, called the *chip rate*, is hundreds or thousands of times faster than the original data, thus expanding the signal bandwidth by the same ratio. CDMA effectively spreads the carrier out over a bandwidth that relates to that of the random bit stream. Because of the bandwidth-expanding property, CDMA is also referred to as spread spectrum modulation. The information can be recovered by multiplying the incoming CDMA signal again by the original high speed chip stream, a process called correlation detection. If the two streams are synchronised correctly, the user data literally pops out. Otherwise, the output is noise-like. Another CDMA approach, called *frequency hopping*, takes an FDMA carrier and moves it randomly across a much wider frequency band.

Figure 13.25 presents some of the applicable characteristics of CDMA. Actually, CDMA is older than TDMA in terms of the art and practice of radio communication. Much of the research and application engineering work on CDMA has been shrouded in secrecy since World War II. This is because CDMA has some particular benefits in military communications and air defence. For example, a signal can be spread so much that it disappears in the ambient noise present in radio receivers. Another 'military' benefit is that a narrowband jammer, that is present say within the band of the CDMA signal, is spread out when mixed with the chip rate in the receiver. In contrast, the spread CDMA signal shrinks back into its original form in the same action, so that the jammer has practically no effect. However, if 100 jammers were within the bandwidth of the CDMA signal, the interference in the above example would be harmful.

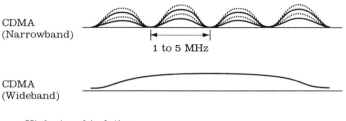

CDMA
(Narrowband)

1 to 5 MHz

CDMA
(Wideband)

• High signal isolation
• Reduced power spectral density
• Critical need for power control

Figure 13.25   CDMA

CDMA is being advocated for terrestrial cellular telephone and satellite voice use over MSS.   This is a significant departure from the limited commercial experience, but is one that raises several interesting possibilities.   Recall the previous comment about TDMA's vulnerability to frequency-selective multipath fading, which forces us to limit the bandwidth of the TDMA signal. CDMA, on the other hand, can live with selective multipath because of the redundancy in the signal. Recall that each bit of user data is shifted around by a stream of hundreds of bits in the chip sequence. It turns out that the correlation receiver can tolerate many missing bits, but still recover the original bit of information.

Obtaining this benefit from CDMA requires that the signal be spread sufficiently to allow selective multi-path to take hold. However, there is a body of evidence that suggests that the amount of spreading is well in excess of what is currently on offer for use on MSS satellites. If there is not sufficient spreading, then multipath will suppress the CDMA signal, just as it does FDMA and TDMA. This is not an impossible situation, since one can compensate with power control and/or additional link margin.

Another benefit of CDMA has to do with its interference rejection properties. FDMA and TDMA rely on cell separation to provide sufficient isolation between transmissions on the same frequency. The amount of isolation depends on the cell pattern and specific link design, but is generally in the range of 9 to 18 dB. CDMA, on the other hand, lives with interference from multiple users because of the separation afforded by the lack of correlation.   In a fully loaded system, interference will limit performance because undesired signals appear as noise. This is an extremely complicated subject, because of the interaction among many design parameters. A properly engineered FDMA or TDMA system can accept beam to beam isolation of 16 dB or less as long as frequencies are assigned intelligently.

Add to this that most of these systems are power or bandwidth limited, and it can be seen that interference may play only a secondary role in system capacity.

In summary, FDMA, TDMA and CDMA each can provide a viable access methodology for terrestrial cellular and satellite mobile systems. There are many tradeoffs in the selection, but there does not appear to be an overriding benefit in one with respect to the others.

13.5 Speech compression

Voice coding and compression is a process whereby speech is converted into a digital bit stream and compressed to reduce the amount of bandwidth needed for transmission.   It is a technique that has literally dozens of alternative implementations, some of which will be reviewed later. The bottom line in coding and compression is that the number of bits per second is cut substantially, yet the listener may notice only a slight impairment in the speech quality or listening effort. The truth is that the speech that is recovered at the receiver has been changed to some degree. On the other hand compression of this type increases the effective capacity of an MSS, or cellular network, by a factor of two to five. Interestingly, much of the improvement claimed by some proponents of CDMA and TDMA is actually the result of coding and compression and not from the multiple access method itself.

Capacity is increased because of the principle that the amount of power needed to transfer a voice signal is proportional to the amount of information per second. In addition, the amount of satellite power required decreases by reducing the number of bits per second being transmitted. This also has the benefit of reducing the amount of bandwidth needed by the satellite.

As time goes on, two improvements are likely to occur. First, the bit rate can be further reduced without a corresponding degradation of quality. This has already been demonstrated in simulation for a 2.4 kbps rate, which represents another doubling of capacity. On the other hand, 4.8 kbps speech can be improved to reach toll phone quality.   That means even fewer people will notice any difference from a normal digital telephone circuit.

Figure 13.26 presents a survey of various voice coding and compression systems.   Speech coders fall into three general categories: waveform codes, vocoders, and hybrid systems. The most popular form of coding though least effective in terms of compression, is the waveform coder. Included is the familiar PCM coding scheme, the standard for all telephone networks. The standard 64 kbps rate of PCM can be halved by employing Adaptive Differential PCM (ADPCM), which is also now adopted as a world standard.   ADPCM is indistinguishable from PCM for voice communication. To achieve a data rate of 16 kbps or less, one needs to filter out non-speech information and compress the rest, which is the approach taken in the vocoder and hybrid systems as well. A vocoder actually substitutes synthesized speech elements for those of the sender.

Hybrid systems do a little of everything, with the intent of producing lifelike speech, which tends to increase the data rate, while compressing much of the data through techniques found in vocoders.

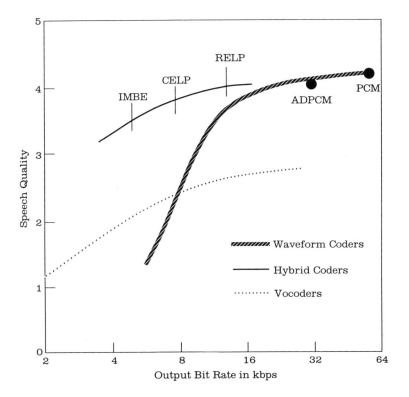

Figure 13.26 Compression system quality and performance

The two most popular speech coders use linear prediction, which requires that a segment of digitised speech be read into a memory for analysis. What results is a compressed datagram, which can be forwarded to the distant end. These work very well in practice, but increase the time delay, due to the requirement to store a segment of speech at the sending and receiving ends. In comparison, the feed-forward technique found in RELP tends to have much less time delay, but at the expense of delivering less compression. Currently, coders that operate around 4.8 kbps impose the memory requirement while those that operate at 16 kbps or more do not. Another consideration is that the greater the complexity of the coder, the more physical circuitry and electrical power that must be accommodated. This tends to work against allowing a very compact hand-held instrument, but is not a problem for portable and vehicular units. The hope is that a new breed of low-delay codecs will deliver acceptable speech quality at 4 kbps or less. This could be critical to the success of MEO and GEO networks, where propagation delay is significant.

13.6 Beam coverage options

The Optus B satellite covers the Australian land mass with a single wide beam. This spreads the available power over a wide area and thereby reduces the radiated power density. The consequence is less signal level, which must be compensated by both a higher gain ground antenna and reduced capacity. This concept is also illustrated in Figure 13.27, which shows an area coverage beam for Asia. With a single beam, the full spectrum is available everywhere and there is no opportunity to reuse frequency. The heavy contour represents the minimum specified EIRP and G/T, and the outer contour shows a wider area with about 3 dB less performance.

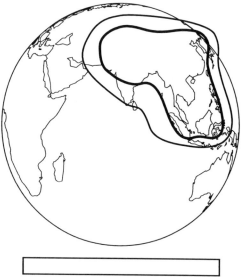

Frequency Allocated to Beam (No Reuse)

Figure 13.27 Spacecraft antenna single beam coverage

A preferred approach for the next generation of MSS satellites is shown in Figure 13.28. With multi-beam coverage, the area is covered by an array of cells, not unlike terrestrial cellular. The allocated spectrum would be divided into seven segments, as indicated at the bottom of Figure 13.28. Beams that are separated by two beams can reuse the same band segment. This particular example provides 100% frequency reuse.

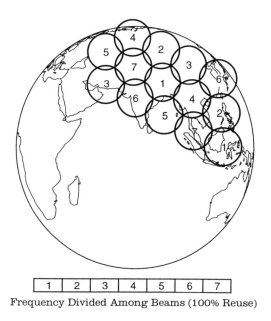

Frequency Divided Among Beams (100% Reuse)

Figure 13.28 Spacecraft antenna multi beam coverage

Another benefit of the smaller beams is that the gain is higher in inverse proportion to the area. In this particular example, the relative gain advantage is approximately 14 to 1, or 11 dB.

This is the technique proposed for all new MSS systems, since it directly benefits the network. To increase the gain, it is necessary to increase the size of the spacecraft antenna. Small beams mean large antennas, but the network capacity goes up and the size of the mobile antenna goes down. These are two very important results.

13.7 Economic comparison of systems

Selection of the orbit geometry and network architecture are major decisions for developers of MSS systems. This section considers the tradeoffs that such developers must go through to arrive at a cost-effective system. Any network, no matter how technically advanced, must meet its rate of return requirement and survive in a competitive environment. Both of these are difficult challenges, particularly considering the number of technical features that must be proven before MSS for hand-held phones becomes a reality.

13.7.1 Methodology

The economics of MSS are evaluated in the context of a global network. It is also assumed that any such network must meet a reasonable financial return (taken here to be 15%). The single most important factor in the cost of the network is the number of satellites required to complete the constellation. Figure 13.29 presents the number of satellites required to complete a global network as a function of the altitude, in miles. Note that the critical design parameter is the altitude.

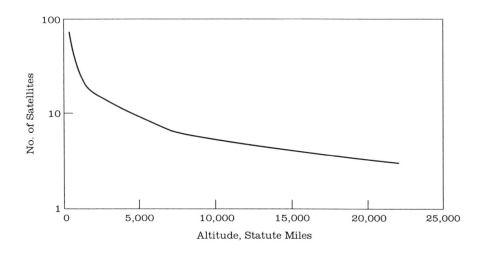

Figure 13.29   Theoretical number of satellites required against satellite altitude

A summary of the investment cost for a global MSS network as a function of altitude is presented in Figure 13.30 and the corresponding revenue requirement in Figure 13.31. The network will be designed and constructed differently, depending on the particular orbit configuration. A number of general assumptions have been made to allow this comparison to be made on a like-to-like basis. Thus, the comparison does not directly apply to the specific proposals made by Motorola, Loral, TRW or Inmarsat. A hypothetical GEO global system is considered as well. One particular advantage of the GEO approach that is not discussed is the ability to implement one region at a time, thus reducing the initial capital investment.

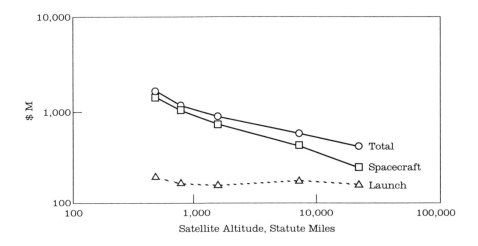

Figure 13.30   Investment cost requirements

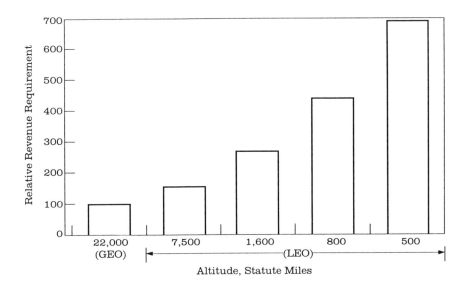

Figure 13.31   Relative revenue requirement for global system space segment

13.7.2 Satellites

LEO systems all must employ lots of satellites. However, the number of satellites required drops rapidly as the altitude is increased. MEO/ICO systems require only 6 to 8 satellites. The lowest number is achieved at GEO, which totals only 3. In this case, GEO does not provide coverage of polar regions. However, because such a small population lives above the Arctic and Antarctic circles, no significant market is lost.

A factor to keep in mind, as the altitude of a satellite increases, is the power required to satisfy the link budget. There is a net disadvantage of approximately 10 dB between LEO and GEO. This results from increased path loss (about 30 dB), compensated for by the increased spacecraft antenna gain due to the use of narrower spot beams on the GEO satellite. Taking this into account, the GEO satellites are somewhat larger and more expensive. However, these factors are more than compensated for by the fact that there are significantly fewer GEO satellites needed, and these satellites can be optimised for coverage of the land masses.

13.7.3 Launching services

Launch services are needed to get the satellites into orbit. LEO systems have the advantage that they require substantially less energy to place each satellite in orbit. As well, it is possible to simultaneously launch several LEO satellites on the same vehicle. For GEO, it is possible to launch at most two satellites at the same time. MEO systems lie somewhere in between.

Regardless of the final altitude, all satellites are first placed in LEO as a staging area. Then additional rocket motors are used, if necessary, to move the satellite into its proper orbit. The LEO satellites only need the first boost and hence can save on additional stages. This reduces the weight that must be launched in the first place. However, as indicated as the bottom curve in Figure 13.30, this advantage is lost by the need for many more satellites in orbit. A conclusion is then that the total launch cost to initiate operation is nearly insensitive to altitude. This does not consider the cost of replacement launches for the LEO systems, for which satellite lifetime is approximately half that of MEO and GEO.

13.7.4 Ground segment (infrastructure)

The ground segment to provide the gateways and network control is different for each of these networks. GEO systems tend to be the simplest to operate for two reasons. First, the satellites are more or less stationary and do not require much attention. Second, there are fewer satellites. The LEO systems will be a substantial challenge to manage. Gateway earth stations must track at least two satellites at a time to ensure continuous service. Links must be handed off, just as in terrestrial cellular, but the time delays and distances involved will make this problem complicated. In this analysis, it is assumed that the ground segment costs are identical and therefore can be excluded from the comparison.

13.7.5 Adding it up

The total investment shown in Figure 13.30 decreases monotonically as altitude increases. The cost of the satellites themselves is proportional to the number of satellites. The GEO satellites are heavier and more expensive than those for the lower altitudes. However, this factor is dwarfed by the quantities of satellites needed for non-GEO.

Adding the cost of satellites to the cost of placing them in orbit, preserves the downward slope as altitude is increased. The conclusion here is that, from an investment standpoint, GEO MSS systems are less expensive than any other altitude.

13.7.6 Revenue requirements

The last relationship to be compared for the different orbit geometries is the revenue requirement for the service.   Figure 13.31 plots the relative revenue requirement for each altitude, where GEO is set equal to 100. As altitude is decreased from GEO, the increase in cost of providing service is very pronounced. The disadvantage of LEO relative to GEO is by a factor of almost 7 to 1. This is driven by the investment cost of initially deploying the satellites, but is pushed further by the fact that the LEO satellites only last about half as long as the GEO and MEO satellites.   This is because drag from air molecules at this altitude eventually causes the satellite to re-enter the atmosphere. Sufficient fuel is needed to counter this force and planned operators are providing for only six years or so (compared with 12 to 15 years for GEO and MEO).

With this type of differential, one might wonder why some of the leading high technology companies have embraced LEO systems.   One reason is that LEO satellite links have the least possible propagation delay, which is advantageous for voice service. On the other hand, there have been several studies which show that the extra delay, to even GEO, has only a slight effect on the acceptability of the service. This is because all MSS systems will use digital voice compression down to 4.8 kbps or less.

This basic comparison illustrates the need to look at the top-level system parameters before choosing a satellite network configuration. The LEO decision is based on the few technical benefits of being close to the earth, but does sacrifice simplicity and cost.   Perhaps the MEO/ICO approaches will prove viable, since their fundamental costs are significantly lower than for LEO. On the other hand, it would appear that the GEO solution continues to be the lowest in overall cost, and is the simplest to implement and operate.

13.8 Conclusions

Satellite based mobile telephone systems are still emerging, as evidenced by the relatively low penetration levels of the service. Still, the promise is exciting because there are countless areas where this type of service will be valuable and probably commercially successful. The major focus now is on how to make the satellite link useful to hand-held phones. This is a major challenge because of the limitations of the simple antenna on the unit, and because of the shortage of viable spectrum. The fact that frequency reuse is so difficult with MSS networks adds to the problem. Still, there are great benefits to the realisation of this dream of ubiquitous MSS with hand-held phones.

Terrestrial cellular networks will probably work better and be less expensive to use in urban environments. The places where satellite networks will shine are likely to be the underdeveloped regions and those 'dead spots' that exist in land-based radio networks. The key to success of MSS will be to find those niches and to exploit them in a way that is complementary to the successful cellular networks of today.

References

1   Elbert, B.R. and Louie, P.A. 'The Economics of Mobile Service by Satellite,' The 14th International Communications Satellite Systems Conference, March 22-26, Washington, D.C., AIAA, 1992.

2   Elbert, B.R. *Introduction to Satellite Communication*, Artech House, Inc., Norwood, MA 02062, 1987.

3   Haugli, H.C., Hart, N., and Poskett, P. 'Inmarsat's Future Personal Communicator System', The International Mobile Satellite Communications Conference, California Institute of Technology, Pasadena, CA, June 17, 1993.

4   Macario, R.C.V., on behalf of The Institution of Electrical Engineers (1991), *Personal and Mobile Radio Systems*, United Kingdom: Peter Peregrinus, Ltd.

5   Wagg, M. 'Regional Satellite Developments - MOBILESAT', Mobile Satellite Communications in Asia Conference, AIC Conferences, Hong Kong, December 6 & 7, 1993.

6   Whalen, D. J., and Churan, G. 'The American Mobile Satellite Corporation Space Segment,' The 14th International Communications Satellite Systems Conference, Part 1, A Collection of Technical Papers, March 22-24, 1992, Washington, D.C., AIAA, 1992.

Index